A Practical Introduction to Supply Chain

OTHER SUPPLY CHAINS BOOKS FROM CRC PRESS

A Guide to IT Contracting: Checklists, Tools, and Techniques
Michael R. Overly and Matthew A. Karlyn
ISBN 978-1-4398-7657-2

Electronically Stored Information: The Complete Guide to Management, Understanding, Acquisition, Storage, Search, and Retrieval, Second Edition
David R. Matthews
ISBN 978-1-4987-3958-0

Enhancing Enterprise Intelligence: Leveraging ERP, CRM, SCM, PLM, BPM, and BI
Vivek Kale
ISBN 978-1-4987-0597-4

Enterprise 2.0: Social Networking Tools to Transform Your Organization
Jessica Keyes
ISBN 978-1-4398-8043-2

Enterprise Integration and Information Architecture: A Systems Perspective on Industrial Information Integration
Li Da Xu
ISBN 978-1-4398-5024-4

Implementing SAP® CRM: The Guide for Business and Technology Managers
Vivek Kale
ISBN 978-1-4822-3142-7

Information Security Fundamentals, Second Edition
Thomas R. Peltier
ISBN 978-1-4398-1062-0

Information Security Policies, Procedures, and Standards: A Practitioner's Reference
Douglas Landoll
ISBN 978-1-4822-4589-9

Information Systems: What Every Business Student Needs to Know
Efrem G. Mallach
ISBN 978-1-4822-2370-5

Leadership in Chaordic Organizations
Beverly G. McCarter and Brian E. White
ISBN 978-1-4200-7417-8

Leading the Internal Audit Function
Lynn Fountain
ISBN 978-1-4987-3042-6

Managing Risk and Security in Outsourcing IT Services: Onshore, Offshore and the Cloud
Frank Siepmann
ISBN 978-1-4398-7909-2

Operational Assessment of IT
Steve Katzman
ISBN 978-1-4987-3768-5

Oracle Business Intelligence and Essbase Solutions Guide
Rosendo Abellera and Lakshman Bulusu
ISBN 978-1-4822-3407-7

Responsive Security: Be Ready to Be Secure
Meng-Chow Kang
ISBN 978-1-4665-8430-3

Securing an IT Organization through Governance, Risk Management, and Audit
Ken E. Sigler and James L. Rainey, III
ISBN 978-1-4987-3731-9

Security without Obscurity: A Guide to Confidentiality, Authentication, and Integrity
J.J. Stapleton
ISBN 978-1-4665-9214-8

Software and Systems Architecture in Action
Raghvinder S. Sangwan
ISBN 978-1-4398-4916-3

Systematic Strategic Planning: A Comprehensive Framework for Implementation, Control, and Evaluation
Hakan Butuner
ISBN 978-1-4987-2481-4

The Art of Agile Practice: A Composite Approach for Projects and Organizations
Bhuvan Unhelkar
ISBN 978-1-4398-5118-0

The Complete Guide to Cybersecurity Risks and Controls
Anne Kohnke, Dan Shoemaker, and Ken E. Sigler
ISBN 978-1-4987-4054-8

The SAP Materials Management Handbook
Ashfaque Ahmed
ISBN 978-1-4665-8162-3

The Strategic CIO: Changing the Dynamics of the Business Enterprise
Philip Weinzimer
ISBN 978-1-4665-6172-4

Web Security: A WhiteHat Perspective
Hanqing Wu and Liz Zhao
ISBN 978-1-4665-9261-2

A Practical Introduction to Supply Chain

David Pheasey

CRC Press
Taylor & Francis Group
Boca Raton London New York

CRC Press is an imprint of the
Taylor & Francis Group, an **informa** business

CRC Press
Taylor & Francis Group
6000 Broken Sound Parkway NW, Suite 300
Boca Raton, FL 33487-2742

Printed on acid-free paper
Version Date: 20160328

International Standard Book Number-13: 978-1-4987-4894-0 (Hardback)

Library of Congress Cataloging-in-Publication Data

Names: Pheasey, David, author.
Title: A practical introduction to supply chain / David Pheasey.
Description: Boca Raton, FL : CRC Press, 2015.
Identifiers: LCCN 2015047454 | ISBN 9781498748940
Subjects: LCSH: Business logistics--Management. | Materials management.
Classification: LCC HD38.5 .P495 2015 | DDC 658.7--dc23
LC record available at http://lccn.loc.gov/2015047454

Visit the Taylor & Francis Web site at
http://www.taylorandfrancis.com

and the CRC Press Web site at
http://www.crcpress.com

Contents

Author

I have worked as a supply chain consultant, much of it on a global basis, for nearly 20 years. This has involved training and consultancy work focused mainly on reducing costs, lead times, inventory, and risks. This work has taken me to Europe, the United States, Russia, Japan, China, and many other parts of the Far East and Middle East. I believe a good supply chain is central to the success of most businesses and remains an area of great potential for most.

I began my working life as a mechanical engineer for the Lucas group in the United Kingdom and I am now a director of Demand Chain International. Manufacturing in some form or another has been a constant ingredient in my working life. It is an area I enjoy greatly. Early work in new product introduction and manufacturing projects led me into the IT world of material requirements planning (MRP) and shop scheduling. This enabled me to move into electronics manufacture and I became head of the supply chain for Research Machines, a company making PCs and writing software. I then learned a lot about cellular manufacture and how to devolve tactical supply chain work to cells and minibusinesses working for Ingersoll Engineers as a supply chain consultant.

Research Machines enabled me to deal with suppliers in both Japan and the United States. I experienced the very different approaches to business of these two cultures. The Japanese approach to quality was also a great influence on Research Machines and on me. The lengthy lead times of electronic components such as DRAMs, EPROMs, PALs, ASICs, and the like were a significant challenge. This, combined with the short customer lead-time expectations emphasized the real value of sales and operations planning (S&OP). We also developed an innovative commodity-based structure that brought focus and responsiveness to the commercial supply chain functions.

In the 1990s, Ingersoll Engineers was a manufacturing consultancy and leading light in the implementation of manufacturing cells and minibusinesses. I joined them in my first consultancy role and was responsible for implementing tactical supply chain work within cells. This was very successful in bringing a closer working relationship between the cells and their suppliers and customers. I managed a number of projects implementing cells and minibusiness structures, including some large businesses where responsiveness to customer requirements and communication with suppliers were transformed.

I left what has now become the Bourton Group to found Demand Chain International with Richard Wardle in 2001, a consultancy focused on improving supply chains. Since then we have delivered a range of successful projects, mostly for businesses with global operations and suppliers. These have often included S&OP, MRP based on clear demand/supply characteristics for each part, contractual and scheduling arrangements with international supply networks, inventory reduction, and Lean improvements in the supply chain. DCI's clients include small and large businesses, each challenging in their own way. Private equity house clients have been highly satisfied with inventory reductions measured in tens of millions and those with Far East supply chains with over 30% reductions in lead time.

I have been fortunate to work with many talented people and I thank them for sharing their time, experience, and friendship with me over many years. I first met Richard Wardle at Research Machines and for more than 20 years he has been consistent in his thinking, hard work, values, and good humor. I have also traveled far and wide across the globe with Crispin Brown. We have eaten in many strange places and found the effects of jet lag do not diminish. His thinking ability and good humor have also got us through some tough times. I also thank my wife Carolyn for holding the fort and putting up with my many absences.

Internationality is increasingly important, especially in manufacturing and supply chain consulting. Demand Chain International is a member of the International Consulting Network (ICN) and I am currently chairman.

DCI is a member of the International Consulting Network Limited.

List of Acronyms

ABC	A, B, C parts management
AGV	Automatic guided vehicle
ASIC	Application-specific integrated circuit
ATP	Available to promise
BFO	Best and final offer
BOM	Bill of material
BPR	Business process reengineering
CAE	Computer-aided engineering (software package)
COGS	Cost of goods sold (accounting term)
COO	Country of origin
CRM	Customer relationship management
CTQ	Critical to quality
DFM	Design for manufacture
DFSC	Design for supply chain
DMAIC	Define measure analyze improve control
DPO	Days purchases outstanding (how much do we owe suppliers)
DSO	Days sales outstanding (how much do our customers owe us)
EAR99	Export Administration Regulations
ECCN	Export Control Classification Number
ECN	Engineering change notification
ERP	Enterprise resource planning
FCPA	Federal Corrupt Practices Act
FIFO	First in first out
FMEA	Failure mode and effect analysis
GTM	Go to market (survey)
IC	Integrated circuit

IPR	Intellectual property right
KP	Kanban population
KPI	Key performance indicator
LLCC	Low labor cost country
MPS	Master production scheduling
MRO	Maintenance repair and overhaul
MRP	Material requirements planning
MRPII/MRP2	Material requirements planning (with feedback loop)
NDA	Nondisclosure agreement
NPD	New product development
NT	New technology
OTIF	On-time in full
P-Card	Purchasing card
PCB	Printed circuit board
PDCA	Plan do check act (Lean term)
PE	Private equity (investment house)
PERT	Program evaluation and review technique
PO	Purchase order
PPV	Purchase price variance
RCCP	Rough cut capacity planning
RFI	Request for information
RFID	Radio frequency identification device
RFP	Request for proposal
RFQ	Request for quotation
ROM	Read-only memory
S&OP	Sales and operations planning
SOX	Sarbanes–Oxley (Act)
SRM	Supplier relationship management
TCC	Trade control and compliance
VMI	Vendor managed inventory
VNA	Very narrow aisle
VSM	Value stream mapping
WEE	Waste and electronics equipment (directive)
WMS	Warehouse management system

Introduction

My colleagues and I are lucky enough to work for a wide variety of international clients where there is a physical product, made in-house or outsourced. In some cases there is some other form of physical operation such as a large plant. In almost all of these businesses, suppliers produce more than half the product content and cost of sales. In some cases this is over 90%. Despite this, we often find the supply chain is undervalued and harbors many of its biggest opportunities left relatively untapped. The reasons for this are not poor purchasing. They are usually rooted in restricted understanding about the supply chain. Old attitudes often prevent us from engaging with suppliers—in case they take advantage of us. Restricted thinking about the supply chain almost results in a boundary around our own organization, beyond which the ability to make innovative developments or sometimes to apply common sense is limited. Some big benefits in cost reduction, higher customer service level, lead-time reduction, and quality are available from suppliers without the need for brilliant theories. Most of the changes are a matter of shaking off old attitudes and using common sense.

In many businesses, even now, supply chain people are trapped in *reactive* roles. They source, contract, purchase, receive, warehouse, and ship as a service. They are often regarded as a back-office function and as such they fail to develop the best contributions suppliers can make. However, in some businesses, suppliers contribute to improvement programs, technology, funding, marketing, logistics, and engineering expertise. Those who work beyond the company boundaries reap the benefit in terms of more competitive costs, quality, and lead times. Suppliers can seldom take advantage of these people—in fact it would not be in their best interests. I remember working for a large technical process industry that selected machining and fabrication suppliers for large engineering programs by obtaining three quotations for each job. We worked as part of their team to help them to select three suppliers to do all of the work on the basis of agreed cost structures and an allocation of capacity. After one of the early monthly meetings, one of the buyers said he was "gob-smacked" by the new cooperative working between the three competitors. One of the suppliers had offered to lend some tooling to one of his competitors. One of the barriers to making the new supplier arrangements was the official management policies. They clearly required three quotations for everything and did not allow for longer-term contracts where suppliers were selected to provide a wider range of capability rather than specific parts. Sensible arrangements for new ways of working can usually be found, while still working within the rules.

Suppliers and the supply chain can contribute in ways that reduce total cost, improve response time, and improve quality, depending on the business. Therefore, there is no ideal contribution, no ideal supply chain structure, no ideal scope or processes, and certainly no ideal system! The best way for the supply chain to contribute to a particular business needs to be worked out with an insight into its strategy, customers, people, and processes. Supply chain people therefore need to be involved in the highest levels of the business. Thinking needs to be innovative in order to maximize contribution to the business. Breaking into a *proactive* supply chain role takes broad thinking, a talent for persuasion, and sometimes the courage to have a go at it.

Having talked a bit about what you might call big thinking about supply chain, this book is about real situations we've seen in our work around a large number of organizations; I haven't made it up. There are not many formulas, new theories, or groundbreaking revelations. I have read too many supply chain books like that. This book is about how to make a supply chain organization work in practice, contributing more to business success than traditional purchasing and logistics organizations can.

In addition to writing about practical supply chain issues and approaches, I have described what we do working with client teams on assignments.

The numbers often surprise us, new issues arise, and things are often not as we thought—which is why you should always *speak with data*. We have also learned that people and their attitudes and behaviors are the most important aspect of making change happen. For this reason I included some of the ways we try to manage the *people* part of the change. I hope you enjoy reading about these practical supply chain approaches and find some of them applicable. My objective is to give every reader something that will make a difference to their own supply chain.

I would like to thank our clients for providing us with so many opportunities to work in their supply chain operations. We have a lot of fun working with client teams, and together have delivered great business benefits.

1

Supply Chain in a Strategic Context

1.1 Importance of Supply Chain Performance to the Business

Supply chain performance has a major impact on businesses in terms of

- Price and cost
- Capacity and lead-time response
- Quality and innovation

We need to have a clear view of how much our supply chains can deliver. We need to be able to persuade senior managers to give us the backing and resources we need to deliver supply chain performance that is a major benefit for the business.

1.1.1 Prices and Cost

Businesses that manufacture, have physical operations, or sell retail products usually spend more than half their cost of sales with external suppliers. The financial impact on the business is therefore clear. For example, in a business making a profit of $50 million and spending $250 million, a mere 1% difference in prices (either way) makes a $2.5 million (5%) difference in profit. The *price paid* to suppliers is therefore important and sometimes very difficult to get right. Changing market conditions, shortages of capacity, changing technology, political issues, and many more factors influence the *right price* to pay. There is no formula or method for calculating the right price. Competitive quotations, comparisons using *price banks*, reference to previous prices paid are all useful. Estimating what you should pay is however usually not a precise science. Judgment, experience, and above all communication and hard work are more likely to get closest.

My colleague Crispin Brown attended a meeting between one of the world's most successful car companies and a supplier. The supplier's quotation was rejected. Their price was too low! The car company knew the price did not include enough money for the level of investment and innovation expected. Driving the price too low can be detrimental in the medium term. You may harm your supply base. Sometimes you may even be deselected by a customer.

In many organizations, price is the most important supply chain issue. It is very easy to measure. Purchase price variance (PPV) (see Section 8.6) is an almost automatic output from the purchase ledger system. However, associated supply costs can often outweigh price differences. Chinese prices and the cost of sea freight from China to the United States or Europe are comparatively low. However, the total lead-times can be 4–6 weeks, and either lack of saleable product or the need to air freight sometimes results. Few people make an allowance for this type of occurrence in their initial cost models.

In addition to the price, there are a wide range of additional costs such as packaging, transportation, insurance, customs, taxes, warranty, and obsolete inventory. Sometimes these can outweigh the difference in price between different suppliers. For example, if you outsource to Chinese suppliers, but lead time to European or United States customers is crucial, inventory holding costs or airfreight may be more than the price advantage.

The competitiveness of most businesses is highly sensitive to prices and supply chain costs.

* * *

1.1.2 Lead Time and Responsiveness

The control of inventory would be easy if customer demand was constant or lead times were instant. We could order the customer requirements from the suppliers as required and have very little inventory. In reality, fluctuating customer orders and varying lead-times make this a challenging area for supply chain. When combined with the fact that the customers usually expect delivery in a much shorter time than our lead time, large mismatches can result. If customer demand fails to materialize, excess inventory levels can result. The larger the excess inventory, the more likely it will become obsolete. Obsolescence costs hurt businesses quite hard. Perhaps even worse, if demand increases too rapidly for supplier capacity, large shortages of product can occur. Customer confidence may never return. The job of the planners is to maintain the right level of inventory to ensure customers still get their product when demand is high, but that the business is not burdened with high inventory costs when demand reduces. At a more strategic level, the supply chain function must ensure the suppliers have sufficient and flexible capacity to meet foreseeable demand. This requires close cooperation between people in purchasing and logistics roles.

The balance between demand and supply lead time, inventory level, and customer service is a constant struggle in most businesses. Complicated processes and systems for forecasting and optimizing inventory are used. It is an important job to make this process as accurate as possible. However, it will always be wrong to some extent. Reducing supplier lead times is a better approach that makes the planning task less challenging. With work, supplier lead-time reductions are realistic, but this type of initiative often receives too little focus. Using our previous example of the $250 million spent by a business making $50 million profit, a 6-week inventory holding represents nearly $29 million of cash. The cost of cash tied up may be an issue, but it is also likely we will need to dispose of some of it due to obsolescence or shelf life. If we scrap 10% of the 6-week holding it will represent a profit reduction of nearly $3 million or 6%. This is therefore a big business issue.

The example of total cost when shipping from China also illustrates how closely purchasing and logistics need to work together. Methods of planning capacity, scheduling, and shipping product need to be agreed early on in the sourcing of parts from overseas (or even from local suppliers). Integrating the organizational structures of purchasing and logistics can also integrate the thinking, and calling them the supply chain works for me.

When I first joined Ingersoll Engineers as a consultant, inventory reduction was a fairly common project. When we started the Demand Chain International consultancy, we helped one of the large private equity firms to reduce inventory in a series of their acquisitions. These were interesting and rewarding programs because they changed the way the planners and the planning system worked. We introduced approaches such as Kanban where appropriate. We changed the way material requirements planning (MRP) parameters were used. We changed the way the planners did their job. We introduced sales and operations planning. Inventory is both a real burden on cash flow and a provider of customer service. In several programs the savings in cash were in the order of $50 million. While the savings are in cash, not profit, they make a business Lean and healthy. Section 5.5.1 describes an approach to "Inventory Reduction and Optimization."

Businesses need to work with suppliers for variable capacity, short lead times, and quality.

* * *

1.1.3 Quality and Innovation

You get what you pay for used to be the view. With quality went higher price. However, in many instances the opposite is true. The "quality costs money" myth was disproved by the Japanese in the 1970s and 1980s when it became clear how continuous improvement was enabling them to produce higher quality at low prices. Now that such a high proportion of our products are outsourced, we are highly reliant on our suppliers and we need to work together with them on continuous improvement. Some quality problems are our fault; for example, poorly defined tolerances on drawings, poorly communicated software updates, material specifications from different regions, even units of measure. (We saw a beautifully machined component on a visit to a Japanese factory. Unfortunately it was 25.4 times too small! The drawings failed to specify the dimensions were in inches not millimeters.)

In more complex quality improvement, data needs to be collected, the failures analyzed, causes diagnosed, and solutions developed and implemented. This requires integrated effort right along the supply chain and back to suppliers. For some items, we need to actively source suppliers who have the right attitudes and quality approach, otherwise we will not achieve the quality levels that customers demand. We once had quite complex quality problems with electronic assemblies from a large supplier who offered to "solve the problem" by adding an extra few products free of charge to every container shipment. This would have more than covered the number of failures, but would not have met the costs of administration, additional costs of shipping replacements to our end customers, testing, and especially the customer dissatisfaction. We refused their offer and sourced the product elsewhere. The new source was at a higher price, but lower cost! A view from a price-driven purchasing point of view would have been to source the lower-priced product.

Many competitive technical issues critical to business success are also rooted in the supply base. Many businesses are now integrators of technology and product supplied by others. Makers of airplanes, computers, mobile phones, automobiles, and many other products rely on key suppliers to develop new technology as well as manufacture the product. Some businesses design key parts of their product in-house, but subcontract manufacture to suppliers. The communication between design and manufacture then becomes limited, and knowledge that would aid manufacture, improve quality, and reduce cost is lost in communication from the manufacturer to the designer. This becomes even more important if the technology involved is developing quickly. Printed circuit board (PCB) assembly is a good example of where good progress has been made on this issue. Most people have outsourced their assembly of PCBs because it is specialized work and requires constant investment of time and money in new techniques and equipment. PCB manufacturers quickly realized the need for design rules. They produced guidelines to help the designers of PCBs with component spacing, heat shadows, and board layout issues. However, in most cases, the design of the PCB layout was not a core competence for the customer. Many businesses therefore decided to keep the design of the circuits in-house, but outsource the design layout of the board. Some businesses, especially those that work with leading and more fashion-oriented electronic products, see their core competence as marketing and are heavily reliant on suppliers for the development of new types of screens, communication technology, and electronics. The degree to which you outsource is a key business decision. Few key players in the fashion electronics market do much of their own manufacturing. However, because they outsource such a significant amount, they are careful to choose a good partner and get the relationship and understanding right.

The value of supplier contribution to engineering, design, marketing information, and quality improvement is difficult to assess. Somehow, the job of the supply chain is to include all the necessary factors into the assessment and management of suppliers—from selection and appraisal to schedule management and even to obsolescence.

Traditional approaches that we still see in many organizations include sourcing on price, relying on the leverage of one supplier against the other, using traditional legal contracts to set-out the business relationship, and keeping suppliers at a safe distance. These traditional approaches won't deliver the best performance from the supply chain. Where old attitudes exist, we need to change them. Our businesses need the most competitive advantage our supply chain can get for them!

Businesses need to work with their key suppliers to develop, design, and supply new technology.

* * *

1.2 Scope of the Supply Chain

Figure 1.1 uses the rather overused supply chain picture, for which I apologize. However, it does show that the number of supply chain issues to be managed along the route to the customer is astounding.

A large range of issues need to be managed along the supply chain. Many of the issues seem to be outside the scope of the supply chain. Forecasting customer priorities and demand patterns, for example, normally belong to sales and marketing, and so they should. However, involvement and understanding of the demand side issues from the supply chain are an essential starting point for good customer service and profitability.

We see many businesses where the supply chain still does not start early enough in the development of new products, does not include all the areas of responsibility it should, or finishes before the customer has been satisfied or suppliers have been paid.

New product designs rely on design engineers to look for the appropriate new technology, but the supply chain should also be looking for these opportunities. Visiting suppliers and technology shows with engineers is a good use of some supply chain time. This is also a way to check what the engineers are doing and make sure they

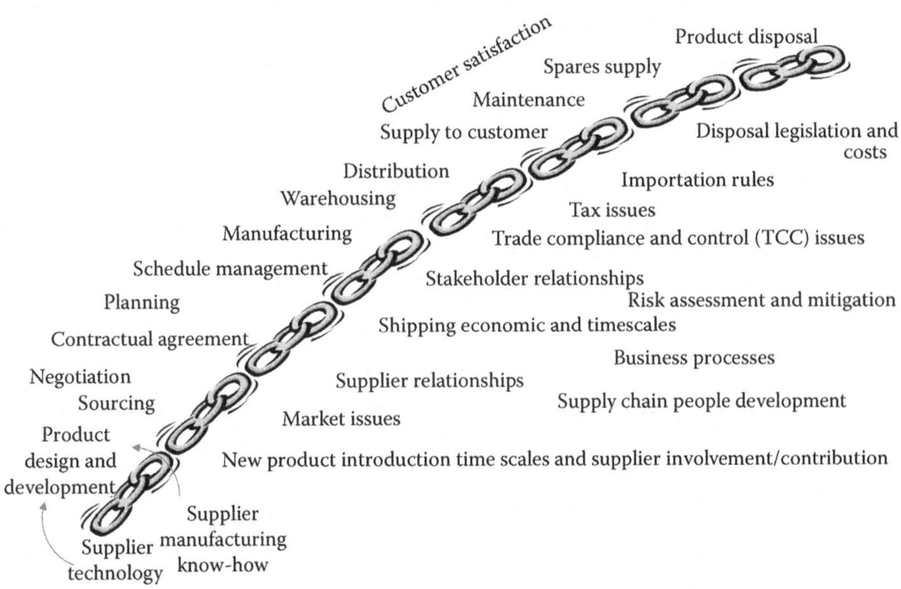

Figure 1.1 Scope of the supply chain.

aren't doing things they shouldn't—for example, making sourcing commitments. This can help to find the best potential new developments in the supply base.

One company we know well designs and markets electronic products. The manufacturing is entirely outsourced—and so too is a large part of the more detailed design work. Experience has proved that the people who will manufacture the product can detail designs that will be more "manufacturable," such as

- Plastic moldings with features, material thicknesses, and draft angles that will work in the mold tools
- PCBs that are efficient to pick and place, work well in reflow, and use components common with other production
- Machined items which use, for example, material specifications that can be sourced from Chinese mills if required, have understandable tolerances and are more easily manufactured using the machine tools available

It is essential that a good process is in place to select the right suppliers early in the product design process. The process is likely to include supply chain people working with sales and marketing as well as engineering and design people.

Some of the best businesses trawl for new suppliers using their *supplier portal.* Potential suppliers can express an interest and after completion of a nondisclosure agreement, they can present their capabilities. Potential suppliers can then be examined to see what they offer.

Working along the supply chain, sales and operations planning (S&OP) is an example of a supply chain process that is of fundamental importance to the business and involves senior people from every function. Examining forecast uncertainties, looking for supply chain options to minimize risks, and trying to optimize the balance of supply and demand is critical and quite difficult. Senior sales and marketing, operations and even finance people need to work carefully together to make the process work. It is one of the most important processes in many businesses, and supply chain usually manages the process and the people.

Beyond S&OP and the subsequent planning activities, a range of logistics work comes into play. Managing inventory and distribution are challenging and costly for many businesses. Customs regulations, taxation, supply of spares, and even disposal of product are all within the scope of the supply chain. Integrated, cross-functional thinking is required through all of these activities.

So how can a supply chain organization contribute to the areas and relationships that are outside its own responsibility? We certainly need to get away from old *departmental* thinking and move to more modern ways of working in cross-functional teams. This is discussed in Section 7.2.

The scope of the supply chain starts earlier, finishes later, and includes more than you think.

* * *

1.3 Supply Chain Organizations and Relationships with Other Functions

A large proportion of the issues that impact supply chain success are outside its scope of management responsibility. Organizing the supply chain to cover such a range of responsibilities and people, needs careful thought and ingenuity. This section is an introduction to thinking about team-based structures. Section 7.2 covers the subject in more detail.

Most businesses still have a functional departmental organizational structure based on principles set out back in the early twentieth century. In their books about *business process engineering*, Michael Hammer and James Champy talk about a fundamental change including organizational structures. The problem is that business processes encounter departmental barriers along their flow. This loses time and focus. Organizing people to do any important task in this way—locating them apart in different rooms, under different management, and with different objectives and measures, would seem to be a strange approach. In *departmental structures*, communication tends to be remote, by e-mail or telephone. It is faster and more accurate to talk to someone who sits near you. When team-based structures are proposed, department heads may say, "what about functional standards, best practice in engineering, finance, marketing, and purchasing?" These are valid arguments because we do have to make sure these specialist tasks are done to the highest standards. However, organizing the whole of the business around specialists does not provide the best solution in our experience. In most cases some kind of *matrix structure* can enable people to be organized in work groups *and* also report to functional heads, but in many cases these become confusing and there is conflict between what is most effective and what the standard calls for. New ways of organizing cross-functional teams are more applicable to the supply chain than any other function.

Others may have used the phrase *organizing around the work*, but I first heard it from Bob Callahan of Ingersoll Engineers in the United States. *Group technology* was the start of this thinking—reorganizing machine shop facilities. Old banks of like machines were replaced by *groups of different* machines, each machine group being able to produce a type of component or product on their own. Cellular manufacturing was developed in which, ultimately the people and equipment in each cell were, as far as possible, capable of running their own business. These reorganizations of resources proved very successful. In fact we are still using the same principles today to achieve better communication, lower costs, shorter lead time, better customer service, and more profit for many organizations.

As a function, the supply chain has the largest challenge in addressing the wider and more strategic issues. Many of the issues are clearly not the direct responsibility of the supply chain; however, the supply chain does need to establish cross-functional teamwork in order to achieve results. New product development is a good example. Engineering teams are usually on tight time scales to deliver new products. Therefore, they need to make the best use of technology available internally and from the supply base to deliver what the customer wants at the right price. To do so, they will need to talk to suppliers a lot about possible solutions, get parts for prototypes quickly, and perform a range of other more subtle and difficult tasks. The supply chain needs to support them in a lot of this work. A starting point is to make sure each design team has a supply chain representative within the team. At least for some of the time, the representative should be physically located with the design team. Their job is to act as a conduit for communication between the design and supply chain teams. They will need to brief people such as senior buyers about technical requirements. They will then need to get involved themselves or set up others within their teams to help to find potential partners who will be best able to help with developments and ultimately products. The supply chain representative will also need to get prototype parts with lightning speed. This will often mean the same day. Purchase requisitions waiting in in-boxes for a day or two is not acceptable.

The supply chain also has to consider issues such as confidentiality, future costing and pricing strategies, how to design for the supply chain (DFSC), meeting the right customer service levels, how suppliers might contribute to the future business strategy, how partnership relationships could work, where we should produce ourselves, and where we should buy.

Chapter 7 (Supply Chain People and Organizations) discusses some of the ways people manage to get the supply chain to work well with other functions.

Supply chain organizations need cross-functional team-based structures.

* * *

1.4 Supply Chain Metrics

We need to be clear about the purpose of performance measurement. The subject is covered in more depth in Section 8.6 (KPIs), but in summary:

1. Some measures monitor situations to check they are in control. These are similar to control limits on manufacturing processes, where the alarm is sounded if the measurements are out of the tolerance range.
2. Some measures drive performance improvement. This is at the heart of continuous improvement and is very powerful for identifying how performance can be improved and not merely monitored (as in point 1).

One of the supply chain difficulties is having to choose only a few measures to focus on; otherwise, the scope is too broad. Section 8.6 covers supply chain metrics that (1) drive improvements and (2) monitor performance. Good measures can drive improvements, but wrong measures not only do not work well, but are even in danger of causing the wrong results.

Be careful to choose the right metrics to drive the change.

***

2

Seven Elements of Supply Chain Excellence

When I had a proper job in supply chain management, I used to think about my organization and where we could improve the job we were doing. Putting together everything necessary for a supply chain organization to contribute well to business performance is a major piece of work. This section describes the seven key elements required for a high-performing supply chain (see Figure 2.1). This also acts as an introduction to many of the areas discussed in detail in later sections.

2.1 Understanding Customers and Demand

The starting point for setting up a business, let alone the supply chain, is to understand your customers and customer demand. For some of our clients we now use a go-to-market (GTM) survey before thinking about significant changes in the supply chain. A recent GTM survey for a client who designs and manufactures a well-known branded product for major retail stores revealed interesting results concerning final delivery points. One of their customers, a well-known U.K. department store, has its own warehousing and home delivery network. They believe (I think correctly) that the quality of this service is part of their selling proposition. However, another of their customers, a large retailer, required increasing deliveries direct from our client to the end customer. The results of the GTM survey led to different inventory levels and distribution strategies (and of course, pricing) for different types of customer demand.

Some supermarkets use their loyalty card system to gather masses of customer spend data through which they hope to understand who shops where, when do they buy, what do they buy, what are the product trends, what is the sensitivity to price/brand, and many other useful pieces of information. Not many businesses can get as much detailed data about their customers as with this method.

Forecasting the volume and timing of customer requirements, although not usually a supply chain accountability, is usually a headache. The sales and operations planning (S&OP) we mentioned briefly in the previous section is covered in depth in the Demand Management section. This process is fundamental to managing the changing balance between demand and supply. Issues such as forecast change, economic cycles, and new product introduction should be discussed and taken into account. One chief operating officer (CEO) told me his S&OP process was a *steering wheel* on the

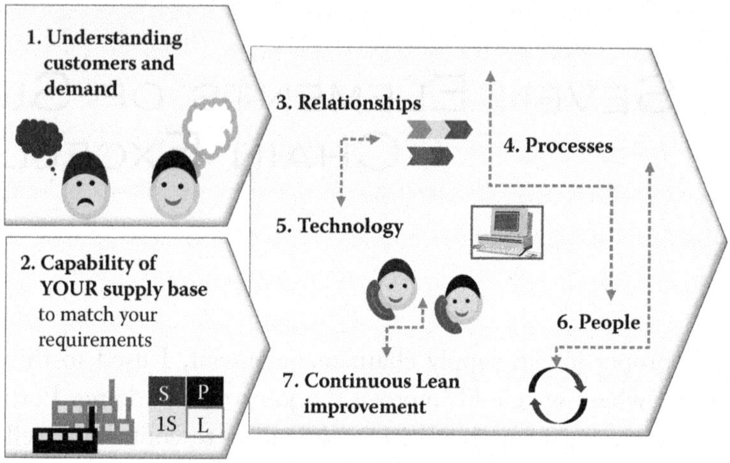

Figure 2.1 Elements essential to excellence of a supply chain with strategic impact.

business. It helped him navigate not only imbalances in demand and supply, but also issues such as changes in product design.

It is also important for the supply chain to understand different types of customer demand. Parts that are required regularly (runners) are easier to manage and should be separated from, for example, parts required infrequently. However, if the infrequently required parts are for urgent repairs then the lead-time must be short. These issues need to be identified and a suitable inventory policy established. Some parts might turn out to be ordered separately, but all at the same time. In this case, should we establish some type of *kit*? This would certainly reduce administrative costs. Supply chain people need to analyze customer demands and segregate different types. The most appropriate way of managing the supply chain for each can then be implemented.

2.2 The Capability of Your Supply Base to Meet Your Requirements

Ideally, suppliers deliver a close match to what our own customers demand. For a tuna canning business, cans need to appear through a hole in the wall as and when the fishes are processed—and of course at the lowest cost, excellent quality, and latest design. For an automotive manufacturer, seats should be delivered to the assembly line only minutes before required, be of the color and type specified by the customer, in sequence, for each car to be produced—and of course, at the lowest cost, excellent quality, and latest type. For a construction project, cement must be available in large and small amounts delivered to exact point of use ready to pour—but at very short notice—and of course be at the lowest cost, excellent quality, and right type. The reason for mentioning these scenarios is that they have all been achieved. However, the supply chain work that has gone into achieving them has involved preparation well ahead of time. This usually requires us to make commitments to our suppliers before we are sure of the demand (at risk). Without this we are unlikely to be able to supply to our customers. The forecasts may be (and are nearly always) wrong and therefore there is risk in doing anything ahead of time. The need for flexibility and the attendant risks may be shared between the supplier and customer, and somehow there must be a level of trust.

The supply base must be efficient in terms of price and cost. There needs to be enough suppliers and capacity, but not too many to make management difficult or to dilute spend with individual suppliers. Small suppliers were seen as problematic for a long time. There were usually a lot of them, the amount of money they represented was small, but the cost of management and administration was disproportionally high. This has changed to some extent. Some businesses advertise the fact that they encourage small, local suppliers. Many are highly responsive and know local conditions. However, administrative costs can be reduced by using P-cards (purchase cards), tiered supply, or buy houses (see Section 4.2). Using this range of practices is now a common way to operate with small suppliers—minimizing the number of purchase orders, receipts, and invoices.

At the other end of the spend curve, the higher spend suppliers need to receive a great deal of attention to make sure prices, costs, and quality are right, capacity is managed, and tomorrow's needs are catered for. This takes a lot of work, communication, planning, and in some cases, shared risk and therefore trust.

For many businesses, suppliers are the major source of technical innovation. We need to select this type of technical partner taking into account the following:

- The way they will be able to interface with our engineering team
- How intellectual property rights (IPR) will be managed
- Supply of similar parts to our competitors
- The possibility that the supplier will become a competitor
- The supplier's ability to continue developing innovative technology

This is the type of supply base alignment we are describing and aiming for. It optimizes the service level provided by suppliers to meet our end customer requirements. Supplier selection is done on behalf of the business, with an evaluated assessment of the risks and costs.

2.3 Relationships

Relationships are what get things done in many business areas. Supply chains involve a wider range of people from different departments and other businesses than any other component. Arrangements are made for deliveries, contracts, costs, engineering changes, warranties, and much more. Some of the time these arrangements may go wrong. When they do, relationships are important because they can often save the day. Although we have contracts that commit people and businesses, it is the relationship which provides confidence that someone will, one hundred percent of the time, do the "right thing." Agreeing on plans for a development or to solve a supply chain issue requires people to sit together in a constructive, trusting, and positively motivated basis.

Almost all the best laid plans of a supply chain go wrong at some point. Not necessarily because they are ill conceived, but because the uncertainties of customer forecasts and changes in real-world circumstances make change inevitable. When things have gone wrong in my previous supply chain jobs, I found that relationships often helped me out. I shudder to think about some of the examples.

On one occasion, we made a mistake in costing a new electronic chip for a new product design. The correct price could have made our product uncompetitive. After long discussions with the European head of the U.S. chip manufacturer who I knew well, he arranged a price concession for the first year. This gave us time to recover the situation.

Another time, we ordered far too many memory chips due to a bill of material error. At the time, chip shortages were easing and prices were tumbling. I went with my purchasing manager to tell the European sales director of the Japanese supplier what happened. I can remember him saying, "I can't believe you're telling me this," but in the end he canceled our excess and probably lost at least any profit on his sale of the resulting inventory.

These people did me a favor. Not because we were weak on pricing or other negotiations; we simply had to be tough because of the competitive market we were in. We had worked hard together with both of the suppliers to improve the way our businesses worked together. Methods of communicating on schedules, technical issues, and even marketing issues had been improved and we had established ourselves as hardworking and trustworthy. This worked both inside the business and outside it with suppliers from Japan and the United States.

Relationships are not just important for problem solving. Relationships forged in the supply chain often result in development and innovation. An informal discussion with a supplier containing an "if only we could" comment can often result in a phone call to say "we think we can do that." A basic example I remember was the desire for a hydraulic pump of a particular type that would continue to pump in the same direction whichever way the shaft was turned. It was only a few days after we had the discussion that the supplier called to say he had modified one of his pumps to do this and he wanted to demonstrate it to us. Examples on a much larger scale have resulted in new factories and even businesses built on relationships.

2.4 Supply Chain Business Processes

Supply chain processes cross more departmental boundaries than any other department. They also cross the boundary into our suppliers' businesses. Getting processes working properly may not seem like exciting work. However, done properly and with the right scope, involvement, and detail, they will usually not only bring good improvements in efficiency, but also provide major benefits in clarity of understanding and better relationships. Our process work has been influenced heavily by working with Japanese suppliers. On occasion, I confess I have been less than enthusiastic about checking through for the third time the procedure to update shipping schedules with Japanese suppliers in a Tokyo hotel when it's long past the time to be sleeping! However, my experience in setting up arrangements in this meticulous Japanese way is that once concluded, things do not tend to go wrong.

Supply chain business processes are discussed in Chapter 6 and process mapping in Section 8.3. Every time we are involved in mapping an existing business process we find surprising things—often additional actions or rules applied to cover a special case that are now part of the standard, but which add cost, delay, and little value. We also find processes that are quite important to the business, but have been developed in a haphazard way and often depend on personal e-mails. These are often fixed quite easily with more defined processes and use of tools such as specially formatted spreadsheets on a shared server.

We were involved in process-mapping a payment process with a problem. Our client was on hold with their supplier and nobody could figure out why. Accounts payable were paying the invoices as fast as they could. Process mapping with the people who did the work uncovered that the invoices for this employee-related service involved the invoices being routed initially to the human resources (HR) department. Nobody in management knew this. It did not take long to fix!

Fixing processes can benefit from help and guidance from senior management, but the people who do the work are the only ones who know what is happening right now, and therefore have to be involved in developing any detailed improvements. Spending some time defining the processes around your supply chain will pay dividends in faster response and fewer miscommunications.

2.5 Information Technology

The supply chain processes could not manage without information technology (IT). Enterprise requirements planning (ERP) systems are now the foundation for managing most businesses, providing the databases, processing, and transactions for almost anyone who has a physical product made in-house or outsourced. There are a wide range of other packages undertaking tasks from design to distribution as well as a range of identification, mobile communication, and scanning devices that are starting to revolutionize supply chain processes. A large proportion of supply chain processes are web-based. They can provide tailored facilities for customer online ordering through to delivery.

Yet many businesses fail to take full advantage of their IT department. As a result, they overburden their people with unnecessary work. There is a tendency to implement systems, not processes. The work is often thought to be complete if the hardware is in place, data is loaded, functionality is checked, and people are shown how to use the system. These implementations often fall short of delivering the best new processes and users feel disappointed with the resulting system. Now that we have such a range of software and devices at our disposal, it is even more important to think process rather than just IT.

ERP systems have been around for many years. Depending on your package provider, these are either very expensive or extremely expensive. They may have been tailored a little or a lot to meet your specific needs. We would find it hard to run our supply chains without these systems, but we do not often use anywhere near their full functionality. The material requirements planning (MRP) functionality for example is quite complex. We really need a high level of practical knowledge about how the system works to use it well. We need to optimize functions such as planning parameters, lead-time offsets, desensitizing action messages, and so forth. This subject is covered in Section 5.3.

The scope and capability of web-based facilities are now encroaching into the territories that have traditionally been covered by these ERP packages. Processes are now often a mixture of web-based, ERP package spreadsheets on shared servers and e-mails. These are often not integrated sufficiently to provide efficient process support. Trying to make one system (usually a system package) cover everything needed by a business is difficult and some packages have become much too complex trying to fulfill this objective.

A friend and former colleague of mine tells me that web-based systems will take over from the current packages in the not-too-distant future. He does run a web development company though.

The use of systems is discussed in Section 6.2 (Supply Chain Information Technology) and MRP in Section 5.3. MRP web-based purchasing, stores systems, distribution analysis and management.

2.6 Supply Chain People

How people act within a business and the way they work with others is what usually drives success. This is even more significant within the supply chain function.

At a senior level, skills in leadership, communication, relationships, persuasiveness, and teamwork are, as usual, of key importance. The challenge is to get everyone to accept, agree, take some responsibility, and take up supply chain developments across different departments and the supply base.

Supply chain people need to have a good understanding of the business and be able to make good judgments within the business context. Buyers of technical products or services need to have at least a reasonable understanding of the technology they are buying—be it in electronics, metal machining, garment manufacture, aerospace structures, construction materials, capital equipment, project management, maintenance, or shipping and transportation.

Many agree that communication is by far the most important skill for supply chain people. What is meant by "communication" may be simply the ability to listen and talk, conveying facts and ideas effectively. This is clearly very important in the supply chain. However, I think communication can also mean the ability to persuade people of your case and gain acceptance of a new idea or situation. This is obviously an important part of successful negotiation. It is also important when convincing people internally and externally to support, for example, a new way of operating, such as the use of cycle counting as a basis for inventory auditing, the use of P cards for low-level purchases, and direct delivery to end-customers' homes rather than the retailer's warehouse. A lot of these types of change need to be made in the supply chain and people from operators to financial controllers need to be convinced of their virtue. Supply chain people need to have the skills necessary to convince people to make changes.

In some organizations, we encounter people who want to hold onto to the way purchasing departments used to be. The senior buyers used to have a style that was gruff or dismissive, and their negotiations focused on *tricking* the other party (*they did not see that one coming!*) or at best applying leverage. Sometimes, helping these people to take on a new, more modern supply chain role is difficult. They sometimes have difficulties with forming partnership-style relationships with some critical suppliers, or taking a broader view of total cost rather than price, for example. New leadership can often help in demonstrating the new behaviors effectively. Mostly, however, the people we've got will convert very well and use their knowledge of the business enthusiastically in their new roles.

2.7 Lean Practice in the Supply Chain

Lean practice is a collection of simple tools that can be used by teams to make (usually small, step by step) improvements to performance. The book *The Machine that Changed the World*, written by James Womack, Daniel T. Jones, and Daniel Roos, began this movement. It was based on the Japanese car manufacturing practices in waste reduction. Lean also includes the quality improvement methods that were started much earlier by Juran and Deming.

Lean tools are highly applicable in the supply chain. Chapter 8 describes some of the most useful tools and gives supply chain examples of their use. I would select the following four of the tools as the most useful in my supply chain experience:

1. *Process mapping* is useful to find out what is going on—what the real process is. It is also useful for managers, particularly when they are moving into new positions.
2. *Continuous improvement cycles* prove as good for improving supply chain performance and quality as those in manufacturing.
3. *Seven (or eight) wastes* are useful to recognize and gain action from a group on the nonvalue-adding activities in their own area.

4. *5S* is always useful for starting up improvements in an area (particularly one such as stores) and making things safe and productive. Individuals and teams can be given responsibility for their own area.

Six Sigma, as initially developed by Motorola and General Electric, involves the use of statistics to achieve quality levels of one in six sigma. The use of statistics is still fundamental to Six Sigma, but most Lean tools have also been incorporated as methods for making performance improvements.

These seven essential elements need to be in place for a complete and successful supply chain.

* * *

3

Understanding Customers and Demand

The term *demand chain* rather than *supply chain* has often been used to emphasize that the focus must be on understanding customer needs. Customers pay for all our work. Real customer needs must be used to drive the supply chain objectives and processes; otherwise, we will continue to do a lot of work that is nonvalue adding, which is contrary to Lean principles. We may also lose customer business! Although it is a truism that the supply chains need to deliver what customers need, many departments work autonomously with too little real information about real requirements and changing priorities. The point of this section is to highlight that accurate customer requirements need to be reflected through the supply chain organization and through to the supplier. In some cases the customer requirements we target are out of date, incomplete, or just incorrect. As part of the starting point for supply chain programs we are sometimes being asked to undertake go-to-market (GTM) surveys to make sure we are targeting the right customer service issues.

GTM surveys enable us to check that we really understand customer requirements rather than assume we know them. For example, findings may alter priorities from say, price, toward lead time, product longevity, or quality. It is surprising how quickly customer preferences change and this knowledge can help us to differentiate our product offering more effectively. We recently did a GTM survey before starting some work on product cost reduction. To our surprise, although the largest customer was, as usual, interested in not paying too much, a more important issue was that the product was officially tested to certain overseas standards, which would help them to expand their export market. The effort used to achieve this was technical, and to quite a large extent, administrative. However, it was a completely different direction than the one expected. It was also much better because it helped to maintain profit margins.

We use the term critical to quality (CTQ) to define more precisely what the customer wants or needs, and the relative importance. As an example, one of our clients supplies a sophisticated set of process equipment. The equipment is expensive and will wear out after about 2 years of use. The customer may replace the equipment with our client's or our competitor's model. Competition in the market place is increasing and our client has been working hard to reduce costs. However, discussions with customers reveal that the installation cost of our equipment is high—quite significant in comparison with the purchase price. Our customer's CTQ issues would favor equipment that lasted longer, even at a higher price, *or* for which the installation costs were lower. If these needs were met, the customer may prefer to pay a higher price for the equipment. This is only one example of why really understanding a customer's needs is important to the supply chain. Others may be around lead times, optional product features, delivery points, maintenance, and so forth.

Be sure to really understand the customer's needs (CTQ issues). They are likely to have a direct impact in the supply chain.

* * *

3.1 Forecasting Demand

Forecasting demand is usually the responsibility of the sales department. The supply chain should not second-guess the forecast when deciding what to make or buy; otherwise, there will be confusion over accountabilities, accusations about the poor forecasting, and recriminations about the resulting breakdown in customer service or excess inventory. As part of the sales and operations planning (S&OP) process it should be a senior management task to review the risks and costs of purchasing and making to the forecast. Although the forecast will, by definition, be wrong, the sales department should be responsible for the accuracy of the forecast and for meeting it with the orders they achieve. Over a period of time, we can improve the accuracy, get a view of how much it might be wrong, and do things to reduce the risks of forecast inaccuracy. Having said that, forecasts that are consistently too accurate lead to the suspicion that sales are aiming too low.

We worked with one company where the sales manager is responsible for the forecast and achieves very accurate sales results each month. In common with many businesses, he collects forecasts from each of his sales people and holds them accountable for achieving their forecasts. Individual bonuses are dependent on the accuracy of sales to forecast as well as the level of sales. This works well, but a concentrated management effort is needed to ensure that forecasts are not too safe. The other slight problem for the supply chain in this particular company is that the product mix sold is often not to forecast, although the total value is.

There are a number of very good forecasting packages available. These systems usually provide a number of forecasting approaches, similar to the examples in Figure 3.1.

All three techniques are in use in a large range of businesses.

Correlated forecasts are in use extensively, linking product demand to economic factors. I am surprised how accurate this approach can be—often achieving figures of well in excess of a 90% correlation. Forecasts for consumer-type goods often use correlations to economic forecasts, but with added inputs of more detailed data such as house-building rates. One of our colleagues once worked for a major provider of capital for equipment purchase. Their forecasting of the demand for capital was based on a number of economic indices. It is extremely important to this organization that the forecasts are accurate and an appropriate amount of money and effort was certainly used. The levels of accuracy were astoundingly high—well over 95%.

Type	Basis	Example
Correlated 	Forecast is based on known data that influences demand	For example, demand or domestic carpets correlates to housing build rate
Historically based 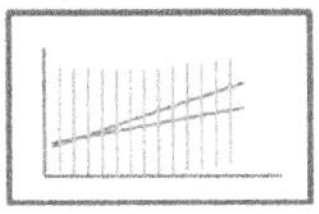	Forecast is based on historical demand that can be adjusted for factors such as growth	For example, next year we forecast each country to sell at the same level as last year, with the same seasonal variations, but with a growth of 10% for countries 1, 2, and 3, and 5% for country 5
Prospected 	Forecast is based on known potential business discussed with customers	For example, businesses selling *projects* to their customers can rate the likelihood of winning each of their opportunities

Figure 3.1 Types of forecasts.

Historically based forecasts basically rely on the past being repeated. They are useful in predicting the pattern of sales through the year and sometimes the mix of product. However, factors of some kind need to be applied to take account of the expected change in business level and product mix. For example, a client selling products into the popular U.K. do-it-yourself market sells a very high percentage of his product during public holiday periods—particularly Easter. The historical forecast is used to predict the annual sales and the seasonal and public holiday peaks. A factor is applied to this historically based forecast based on a correlation with house-building data. The resulting forecasts are usually over 95% accurate. The U.K. weather during public holidays is the biggest unknown factor!

Prospected forecasts are more often found where the number of customers is fewer—for example in capital product sales. Salespeople may work on a *sales pipeline* that may last for several months. Under these circumstances the salespeople often qualify their leads, which can then be used by the forecasting process to predict what percentage of the pipeline will achieve sales. The pipeline concept is a very useful sales tool. It is used extensively in many of our client businesses to assess the likelihood of achieving each sale and also to move actions forward to increase the chances of success. This is our forecasting approach for the consulting business.

One client uses a forecasting package to assess the global demand for a range of high-quality luxury goods. Sales were to some extent seasonal and some products were fashion items with new ranges introduced each season. Sales were made through a mixture of their own brand of retail shops and other large retailers, such as department stores. The forecast was split into geographic regions and some large customers provided their own specific forecasts. Salespeople were also responsible for forecasting their own sales to their retail customers. They used a *customer relationship management* (CRM) tool to collect information from customers, including promotional plans and changes in forecast demand. However, the forecasting system was also able to build a forecast using historical sales data. Some *alpha factor* adjustments were made for economic and other influences. A forecast manager collected all these forecasts and compared the historically based and prospected results for regions and key products. Significant differences, particularly between the historically based forecast and those for specific retailers, were discussed and a *best* forecast was agreed on. Using a combination of forecasting techniques and working hard on correlation with known data, some clients achieve over 95% accuracy with their forecasts.

In the example above, the sales were forecast for the date on which the customer would buy them. However, the supply chain lead time was several months long (typically 3 months). The forecast that is relevant to the sales being made at any date is therefore the one made 3 months ago. This was the point at which they committed to manufacture. This concept should be applied when measuring forecast accuracy (i.e., to be relevant to the supply chain, the current month's sales must be compared with what was forecast 3 months ago).

Forecasts will be used as the main input to the S&OP process, which will commit quantities to be bought and made. A major part of the process is risk assessment and any assessment of forecast accuracy risks will be useful. Good examples of this would be an increase in forecast that is dependent on winning a large new customer or us releasing our new product in advance of our competitors. These risks are real and need to be taken into account by senior management when agreeing to plans for manufacture and purchasing.

Forecasts will, by definition be wrong. However, good supporting data and the use of correlating statistics can produce surprisingly accurate results.

* * *

3.2 Sales and Operations Planning (S&OP)

The importance of S&OP to businesses is easy to underestimate. It is the process that enables senior management to take risk decisions about the balance of supply with demand. One CEO described it to us as like a steering wheel enabling him to manage a key part of his business, matching output to customer demand.

We *do* need computers to do the detailed work, but more importantly, we need an effective business process. The important features of this are

1. *Ownership and senior management engagement.* For the S&OP process to be given the appropriate level of importance it *must* be owned and driven by the CEO/managing director (MD) of the business. His or her team must also include those responsible for both selling and output. They need to be able to explain their situation to the senior team who review and make the key decisions. The sales, operations, supply chain, engineering, HR, and finance departments all have an important part to play in S&OP and should attend the review meeting. Input from each of these areas can influence results.

2. *Highly structured, formal, and rigorous process.* Organizations that maximize the value of the S&OP process treat it with the same level of professionalism they would treat month-end and year-end accounts. There is a fixed timetable that is always and religiously maintained. All input documents are clearly defined including circulations and timings, pre- and post-S&OP meeting are scheduled to maintain consistency of purpose, and actions and decisions are fully documented and available to all. The CEO chairs the meeting the majority of the time and a deputy is appointed for the infrequent times that he or she cannot be there.

   Examples of critical input documents include revised sales forecasts, which need to be available on time. This must be synchronized with the current sales data in order to avoid double counting (see Section 3.2.1 Master Production Scheduling Process). The sales forecast should include some assessment of the accuracy risks (e.g., a part of the volume forecast may be dependent on us winning a particular new customer). This is not to denigrate the forecast—it will be wrong by definition—but some real knowledge about the risks is valuable within the process. The manufacturing/procurement plan also needs to be synchronized to avoid a product that has already been shipped being counted as available.
3. *Ability to determine the best achievable supply plan to meet changing customer demands.* This is usually the role of a master scheduler or sales and operations planning manager. They will need to determine how quickly the manufacturing and procurement plan can be changed—and by how much. This requires understanding of current lead times, capacities, and bottlenecks. This is a role for a highly numerate person who is familiar with the business supply chain.
4. *Ability to communicate and enact our plans quickly following the S&OP meeting.* Manufacturing and procurement plans need to be communicated quickly, usually through the MRP system. The consequences for sales and for customer service levels are also important to be understood and communicated. For example, there may be insufficient production volume available to meet current customer lead times. Generally, it is better to tell customers this before they experience late delivery for themselves.

The S&OP process needs to enable the business to optimize customer service, inventories, and supply chain resources. The impact on customers, cash, internal manufacturing, and supplier performance can significantly impact business performance. The S&OP process will also connect management to what is planned and what is going on in one of the most important aspects of the business—the balance of supply with demand.

The terms *master scheduling* and *sales and operations planning* are often used to mean the same thing. In truth, it depends on who you are talking to. For simplicity and clarity, I will use the term master production scheduling for the (usually) monthly adjustments of the plan to meet changing customer demand. Sales and operations planning will be used to describe the whole process, which also includes the strategic planning of sales and supply. Figure 3.2 summarizes these processes.

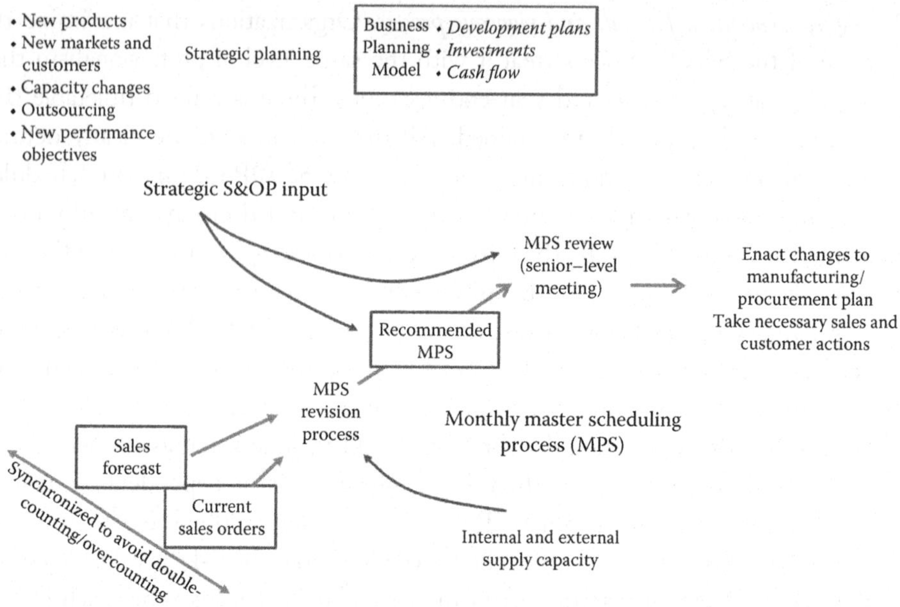

Figure 3.2 Sales and operations planning.

S&OP enables management to evaluate and balance the risks of forecast errors, poor customer service, excess inventory, and inefficient use of resources. The S&OP chairman should be someone senior and often the managing director or chief operating officer takes on this role. Attendance at the meeting and ability to guide the decision making is a major advantage for many managers leading the businesses. Many forecasts have uncertainties (e.g., "if we win this new customer, the trials are successful"). It is useful to have these details to discuss in the S&OP meeting. Often additional revenue can be won—but only if we have the product available, which means taking a risk by ordering some of the materials or reserving capacity now. This is a valuable discussion to have in this meeting, one that hopefully leads to a successful decision on whether to take the risk. This is one reason the finance director or controller should be present because large amounts of cash, costs, revenue, and profit can be at stake. Following approval from the S&OP meeting, the plan can be published for use by both planners and salespeople. Planners will be using the plan to adjust schedules, probably using some form of an MRP tool. Salespeople may well need to use the plan and subsequent updates to provide customers with delivery date information, either on an individual product and project basis, or by updating the general lead time information by product.

It is important to keep the timing tight; one business we know measured the number of days late the S&OP ran on the standard timetable (Figure 3.3).

In this case, the S&OP data was reviewed using a spreadsheet. The forecast also came from a spreadsheet. Existing sales order and inventory data were downloaded semiautomatically from the ERP package. The S&OP revision, which included discussions with sales, manufacturing, and supply chain took 3–4 days, after which the

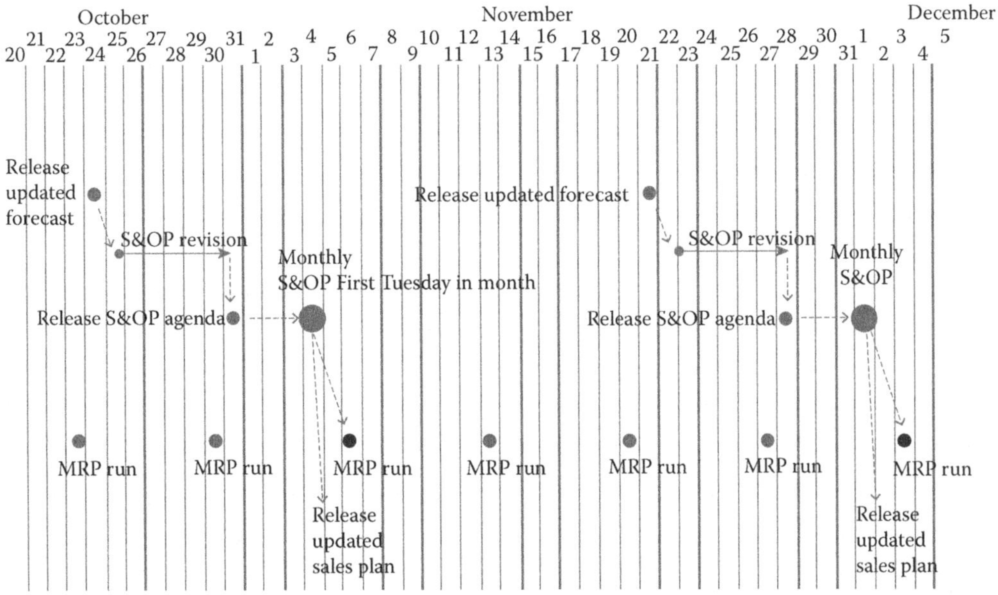

Figure 3.3 Example S&OP timetable.

monthly S&OP agenda and pack were released. Following the monthly S&OP meeting, updated production and sales plans were released. The sales plan usually reported any difficulties such as shortages of supply and set selling targets. The production plan semiautomatically updated the ERP system to drive functions such as MRP.

This example indicates some of the disadvantages of using spreadsheets rather than an integrated systems package. The delays in the process caused by data transfer do cause difficulties. Data errors and misunderstandings also take place and discipline on file versions and backups needs to be rigorous. S&OP functions that are part of the ERP package are still not perfect, still need updating with discipline, may not work in precisely the way you want, and can still be heavy on administration—and cost a lot of money.

Strategic input to S&OP from the business planning and modeling work typically includes changes such as

- Major sales initiatives, new territories, and new sectors.
- New product introductions and phase-outs.
- Changes in manufacturing and the supply chain resulting in capacity and lead time change.
- New costs and prices also need to be factored into the S&OP system. These clearly change throughout the year and some sort of key adjustments to plans are usually needed every 2 or 3 months.

It makes sense for the same senior people who are making the strategic decisions to be part of the monthly S&OP meetings.

Balancing supply with demand is a demanding process involving almost all functions and senior managers.

As indicated by Figure 3.2, the S&OP and business planning model data need to be synchronised. For example, if the business planning model revenue is not in line with the S&OP planned shipments, we are unlikely to meet our business targets. Attendees should be the most senior people responsible for the business unit, particularly in terms of sales, operations, supply chain, finance, and possibly the chief executive. Figure 3.4 shows examples of the roles these attendees could play at the S&OP meeting.

We know a new CEO who joined a high-technology business that was enjoying rapid growth. One of the most difficult business issues was promising delivery dates to customers for the rapidly increasing level of orders. He went along to the S&OP review meeting which his chief operating officer (COO) had just introduced and immediately realized the importance of the meeting and the S&OP process. He decided to be a permanent member of the meeting—in fact, he decided to chair it. This process enabled the company to manage the balance of supply and demand, and sales were able to provide customers with dependable delivery dates.

When MRP systems were first introduced, materials were often planned for manufacturing quantities that could not be achieved in practice. The development of MRP2 and S&OP processes enabled the MRP systems to calculate the details for an achievable plan. Books from the 1980s on the subject are certainly still worth reading. Arguably the best of these are by the late Oliver Wight. Many senior management teams now see S&OP as playing a significant role in bringing about strategic plans and also keeping the ongoing supply and demand under control.

Providers of ERP systems all include some form of S&OP functionality. However, these systems will not ensure success. The business process is far more important. We have experienced some very successful S&OP processes that use spreadsheets for the central part of the master production schedule (MPS) calculations.

CEO	Chair meeting, handing over to the Master Scheduler to explain the demand/supply situation and the recommendations.
Sales Director	Explain any substantial changes in forecast. Press the case for measures which increase customer service.
Operations Director	Ensure the production plan is practically achievable. Press for measures which improve efficiency in manufacturing.
Supply Chain Director	Ensure the production plan is practically achievable in the supply chain. Press for measures which improve the efficiency of the supply chain.
Finance Director	Ensure financial implications of recommendations are understood—revenue, profit, inventory.
Master Scheduler/ S&OP Manager	Produce a summary of recommendations which optimizes the supply chain and manufacturing response to the changing customer demands. Cost any implications on revenue, profit, inventory or other costs such as safety stocks. Explain the recommendations to the meeting according to the agenda.

Figure 3.4 S&OP roles.

The S&OP process also coordinates developments from all other areas of the business that will affect the balance of supply and demand. The following issues therefore need to be on the agenda:

- Product introductions and phase-outs. The S&OP will be managing the ramp-down volumes of the old product and the ramp-up of the new. Slippage of new product introduction dates is a common critical issue. Many companies have developed their own approach to risk reduction. For example, the phased introduction of a new product across different global regions mitigates the risk if the program slips or the product has technical problems following launch. Even on a smooth product changeover, the S&OP coordinates the actions of engineering, operations, and sales.
- Supply chain capabilities, for example capacity during an increase in volume, technical change, or new product introduction are likely to be key to how fast things can be done. Changeover from one contractor to another can be almost as complex as a new product launch, making it another S&OP issue.
- Internal manufacturing changes such as new equipment, shutdowns, shift pattern changes, performance improvements, and anything else that significantly impacts output will need to be taken into account and steered by S&OP.
- Sales and marketing will be planning promotions, entry to new sectors and regions, new customer campaigns, and other strategic actions. It will be counterproductive if the product is not available to meet the new demand. Any new sales may be lost for a long time because customer service is poor.
- Human resources may also be planning training events, considering holiday days, or be part of shift alteration plans. People are usually a critical resource and their capacity needs to be taken into account and driven to some extent by S&OP.
- Larger groups operate S&OP at plant *and* group level. Issues such as the demand/supply balance in different regions can be reviewed by a group-level S&OP and production moved between factories. Tooling may be moved, new product introductions can be switched, and investment plans changed to support meeting changing demand patterns.

S&OP is the process that engages senior managers in all areas to make business decisions based on the latest supply and demand information.

* * *

3.2.1 Master Production Scheduling Process

Figure 3.5 shows a simple example of the master scheduling task for a product being assembled by a business using supplier-provided components.

Units		Apr	May	Jun	Jul	Aug	Sep	Oct	Nov	Dec
Rack 90 system										
Customers orders		22	15		10		20	20	20	
Forecast		92	63	77	80	87	92	110	128	92
Inventory	121	97	109	112	102	95	63	13	-55	-67
Production plan		90	90	80	80	80	80	80	80	80

Figure 3.5 Simple master scheduling spreadsheet.

The example is for a product called the Rack 90 System. The data in all four lines is in units. In this simple case the master scheduler or S&OP manager can balance the production plan in order to meet the latest forecast level of demand.

The total demand is the addition of actual orders and forecast. It is therefore important to *consume* the forecast (i.e., reduce the forecast as orders are taken in a particular time period), otherwise, there will be *double counting.*

The projected inventory line represents the balance between demand and supply. In this example, the business sells from stock and there is a target level for inventory. Some businesses *make to order* and in this case an *order book* line can be used instead (this is the equivalent of negative inventory). It is often useful to show the monthly line balances in value rather than units, for example at sales price for the unit. We need to be clear what unit value is being used, otherwise it can cause confusion. For example, actual sales will probably be reported at the selling prices achieved. This is likely to include some form of discounts. If the S&OP plans are valued at a catalog or even an average selling price, there will be a discrepancy and corrections will need to be made.

The starting inventory of 121 units is projected into each new month end by deducting the total demand (customer orders plus forecast) and adding the planned production during the period. In this case, if the target inventory was 0.5 × monthly demand, planned production needs to be reduced in the short term (Apr to Sep), but then increased. The S&OP function in most ERP systems will provide a new recommended production plan using some inventory targeting parameters. There may be restrictions on which time periods can be amended and which cannot (similar to MRP). However, in my experience, a good master production scheduler who is in contact with the key elements of the supply chain, manufacturing, and customer order management can make more informed decisions. The projected inventory line shows we should decrease production in the short term. Production would then need to increase before the last quarter. However, a thinking master scheduler will question the likelihood of the demand increase in the last quarter and may flag it as a *risk* in the monthly review meeting. There is also a question over the reduction of output in the short term. Should it be decreased to reduce the high inventory in May–July? It may already be too late to reduce supply of some of the high-cost components and the excess inventory would still exist in component form. In fact, there appears to be a *level loading* policy of some sort and it may be more sensible to continue the 90/month schedule rather than reduce it to 80. These are the types of considerations the master production scheduler should be making in order to *optimize response to customer demand.*

Many businesses have rules that govern what changes can be made within certain periods, such as

- No change allowed during current month
- Only decreases or up to 10% increases in month 2
- Maximum of 20% increases in each subsequent month

These guidelines are good, but good master scheduling will be able to make more innovative arrangements to enable customer service to be improved further. These need to be approved by senior management at the monthly S&OP meeting.

The master scheduler (or S&OP manager) circulates a report to all attendees of the *S&OP review and approval meeting*. This generally sets out the following types of information:

1. Significant changes in demand, increases and decreases, and forecast or actual customer orders, why they have arisen, and any levels of risk associated with them (e.g., new customer forecast)
2. Recommended supply response, including the key supply restrictions and issues
3. Summary of the effect of the new S&OP (e.g., impact on customer service, revenue or inventory, additional costs)
4. The agenda for the meeting
5. Summary S&OP data

Before recommending courses of action to the S&OP meeting the master scheduler will need to do some checking. He or she will need to talk to the people responsible for the forecast. Unexpected changes in forecast volume need to be questioned and understood. The forecast should not be changed or second-guessed by the master scheduler. The forecast is the forecast. However, the reasons for significant forecast changes and underlying risks need to be communicated clearly during the S&OP meeting. Key suppliers and manufacturing management will need to be consulted, particularly over any requirement for urgent increases in supply. In many cases, suppliers can beat published lead times and additional capacity can be found. Sometimes these may be at a premium and if large enough, the decision to spend additional money in order to meet possible increases should be discussed at the S&OP meeting. The master production scheduler's job therefore requires good numeracy, experience with the supply chain and operations, and good communication skills.

3.2.2 Available to Promise Calculations

Many businesses need to be able to tell customers when they can expect to receive their order. An *available to promise* (ATP) calculation can provide the information on which to base this *promise*. The ATP quantity for a product next week for example

	Current week	Current week +1	Current week +2	Current week +3	Current week +4	Current week +5
Output available	650	650	700	800	800	250
Projected inventory	50	20	0	90	180	-300
Sales orders	148	392	25	300	210	0
Forecast	600	680	720	710	710	730
Consumed forecast	452	288	695	410	500	730
Available to promise (ATP)	502	258	675	1090	590	250

Figure 3.6 ATP calculation.

could simply be the quantity in inventory, plus products that will be completed, less the quantity already sold in the period. However, if more have already been sold in the previous week than the total available, we cannot promise all of those available in the next week—some are needed to make up the shortfall from the previous week. True ATP calculations therefore start with the period at the end of the product lead time (current date plus lead-time) and work back to the current date period by period. Figure 3.6 illustrates this point. The more comprehensive ERP packages include this type of ATP calculation.

Note that the projected inventory takes into account the forecast, sales orders, and output when projecting the previous period's inventory forward. It predicts what the availability of product will be, assuming forecast sales and output go to plan. In this case there is a significant shortage of the product in *Current week +5*.

The ATP line is completely different. It indicates the product available to meet a customer order now, taking account of sales orders taken, but ignoring any forecast orders. The calculation begins in *Current week +5* and works back in time. (*Current week +5* represents the lead-time of the product). Note that in *Current week +5*, there is a negative ATP and this negative is passed back to the calculation in *Current week +4*.

Beware that some ERP packages calculate ATP as part of the S&OP, but do not update the S&OP file as customer orders are received. This ATP figure does not agree with the ATP that is recalculated by the same package for sales purposes. The ATP figure is updated every time an order is taken.

The next section is about S&OP for products with optional features. Note that some packages enable ATP to be calculated for the optional features.

3.2.3 S&OP for Products with Optional Product Features

If there are only a small number of product options, it is worth scheduling each of them separately. Alternatively, if a large part of the product is standard and there

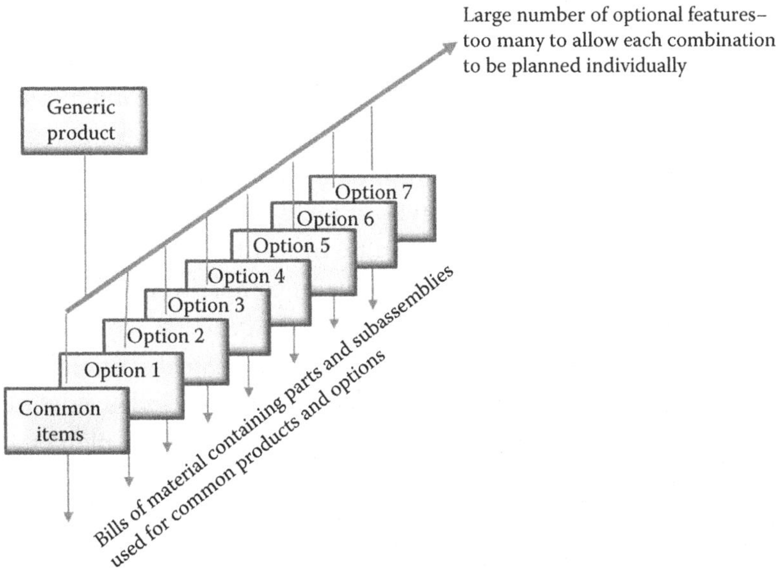

Figure 3.7 Optional product BOM.

are a number of low-cost, short lead time options, it may be practical to include the options as percentages on the bill of material (BOM) as shown in Figure 3.7. In the simplest situation, the master scheduling process operates at the generic product level. Planning for the options to be manufactured is calculated by MRP using the quantity per product of less than one for each option. It is usual to base the quantity per product for each option on the expected use of the option plus some safety. So for an option used 60% of the time, it may be decided to plan at 0.65% per product. There is a problem with this approach because it tends to produce *push out* MRP action messages as the overplanned options are not taken up. The use of a safety stock for these items does not cause this problem, but the safety stocks need to be updated carefully, not only when option take-up changes, but also if volumes increase or decrease.

For complex and costly product options, some packages offer functionality for aggregating and disaggregating between the options and generic product levels. This enables current customer orders and availability at the option level to be aggregated and replanned at the generic level. The plans are then disaggregated down to option level to drive MRP. This is quite complex and differs slightly depending on the ERP package used.

3.3 Dependent and Independent (Spares/Maintenance) Demand

Dependent demand is usually planed and managed using a combination of MRP and perhaps fine-tuned with a kanban or vendor managed inventory (VMI) technique. Independent demand—typically spares, replacement, or maintenance parts—are often less well managed. Demand is really dependent on something of course (failure rates for example), but nothing as simple as a product manufacturing schedule.

Provision of independent demand parts can be managed by looking at historical demand, applying trends, and keeping suitable stocks in place. In practice, this can be difficult because of new product introductions and technical changes or during large changes in product shipment volumes. A lot of product, customer, and historical knowledge is required to achieve good results. In reality, the whole area of independent demand is often relegated to second place in comparison with the mainstream of production requirements. In many instances, independent demand is catered for poorly—from safety stocks, to borrowing from assembly stocks, or adding an allowance on top of production requirements. As a result, when independent demand is critical or urgent, parts often have to be specially expedited in low volumes, generating a lot of supply chain work as well as a poor customer service level.

For many businesses, spare parts represent good revenue and profits. Automotive component manufacturers have historically charged much more for spares than the parts they sell for new vehicles. Aerospace components suppliers often do the same. As a result, spares provision is managed accurately and proactively. Even if the amount of profit from independent component sales is low, we should still apply the best approaches we can, because poor customer service could have a detrimental effect on the business. Supply chain administrative costs could also be high.

In aerospace, there are systems to monitor the service hours and number of takeoffs and landings, as well as the age of key aircraft parts and assemblies. The maintenance of both the aircraft and its key parts is planned using these systems and the kits of parts necessary are supplied ahead of time to the location of the work to be done. These are planned in a similar way to mainstream production. The use of kits is particularly efficient compared with the supply of individual parts and kits could be much more widely used in other industries. Spares for part failure are more difficult, not only because failures are less predictable, but because they can occur almost anywhere. Some spares are held by the airlines, some by the original equipment manufacturers, and a substantial amount is subcontracted to specialist aerospace logistics companies. There is a clear agreement about who is responsible for which spares and where inventory will be held to provide breakdown cover within agreed times to all airports in the network.

The sharing of automotive spares often impresses me. When a main dealer doesn't have what you need, his or her IT system can provide up-to-date information about where the part is in stock at another local garage. This capability has been around for many years and it demonstrates what can be done—even if it means cooperating with competitor dealers!

I had a job where I was responsible for spare parts availability for a wide range of electronic components. I used a special spares bill of material (BOM) to plan how many spares to produce or buy. A special quarterly report showed exactly how many had been used for spares by geographic area. New spare parts were added following new product introductions and requirements were adjusted. Adjustments to the BOM changed the level of spares provisioned through the normal MRP system. I found a particular problem that occurred when maintenance areas replenished their stock using reorder point systems. These systems place single-point demands to replace inventory that had been used steadily over the previous weeks and months. Maintenance engineers restocking their vans caused a similar problem. I found no one solution to these issues, but several things helped:

- More spares kits, which reduced the number of parts to be controlled
- Reorder point systems were not allowed for A and B class items
- Holding more inventory of C class (low value) parts

Some of the IT system packages now offer functionality that can plan and provision parts based on statistical analysis of usage. This function tends to operate at the part level although kits can also be managed. This does work, but we need to tell the system about technical issues, design changes, and similar occurrences that will affect future demand.

For very cheap parts, I worry about the cost of too many transactions in comparison with the part value. Putting low-cost parts into kits is a great help. Oversupply them in the kit if it makes sense; many people I know add extra screws, nuts, and similar items that are very low in cost but are prone to being lost during the maintenance job. Some people I met recently have standard boxes of useful fasteners and other similar C class parts for each of their products. Having these in a kit box reduces the number of transactions needed to get very low value parts. It sounds wasteful, but the cost of having too many is much less than the cost of not having the part you need *or* having to obtain a cheap part at high administrative cost.

Kits are a good way to supply maintenance parts for a number of other reasons. In addition to the lower administrative cost, they are a good way to ensure you have all the parts. A box can be arranged with inserts and subboxes containing individual parts. This makes it easy to visually check that all the components are there and to find them when you need them. In many cases, the maintenance engineer takes the kit box and uses whatever parts are needed to do the work and returns the remainder in the box. This can include tools as well as components. The box can be refurbished and restocked by a supplier or done in-house. Perhaps the ultimate spares kit I ever heard of is a repair/replacement kit used by a leading U.K. airline maintenance function. The kit is supplied to the aircraft engineer who, for example, needs to fix a problem reported by the incoming flight crew. The kit contains

1. All the parts needed (perhaps a main assembly plus other parts)
2. Special tools needed for the job
3. Full instructions on how to replace the part and prepare it for use
4. Provision to collect diagnostic data that is needed about the nature of the fault
   a. Where it occurred
   b. In what circumstances
   c. If the fault was detected or no fault was found

The instructions are supplied and data collection is carried out using the aircraft engineer's computer.

Your solution to managing spares provision may be similar to one of these examples, or you may have to think of new approaches. I encourage you to try to find solutions that match the issues and level of importance of spares in your business, rather than as an add-on to new product provisioning.

Recognize dependent demand where possible and reduce the planning and control workload.

* * *

3.4 Regularity of Demand

Steady, regular demand is much easier to manage than that which is volatile and irregular. This obvious point has been used by consultants for some time, and is often called runner/repeater/stranger analysis.

Stranger requirements are very difficult to forecast. If they are rare spare parts, their criticality can be assessed and a strategic stock can be held if the consequence of failure is high. For other noncritical services, materials, or equipment, we need to allow the full procurement lead time or make some kind of arrangement with providers to reduce the effective lead-time.

Repeater requirements are less risky to hold in stock, and if customers are likely to want the product quickly, forecasting is usually a reasonable approach. Repeaters are usually attributable to some pattern—seasonality, use by a particular customer, repeated project-based demand, and so forth. If the underlying cause of the demand pattern can be found, forecasting and inventory holding will be less risky.

Runner demand is often not worth forecasting, especially if the parts are cheap. Some form of replenishment technique may be better and cheaper. This could involve the supplier or distributor monitoring inventory and replenishment when needed. If a flow arrangement can be set up in combination with some sort of vendor managed inventory approach (VMI, e.g., tell a supplier to expect to supply a certain number every day or week, but adjust deliveries according to how many you have used), the effective lead time will be zero and the risk of excess inventory low.

Parts are easily identified as runners, repeaters, or strangers if you have, for example, details of the historic usage. Strip the usage data and arrange the parts down the left-hand column of a spreadsheet, and usage data in columns for each time period (e.g., weeks). Then add a column to the right of usage to calculate for each usage line; the standard deviation divided by the average (STDEV/AVERAGE). As shown in Figure 3.8, the resulting *coefficient of regularity* (now adopted as a Six Sigma tool) defines the part type.

The reason for doing all of this is to determine how individual parts need to be managed. The techniques for managing runners are easy—arrange some form of flow;

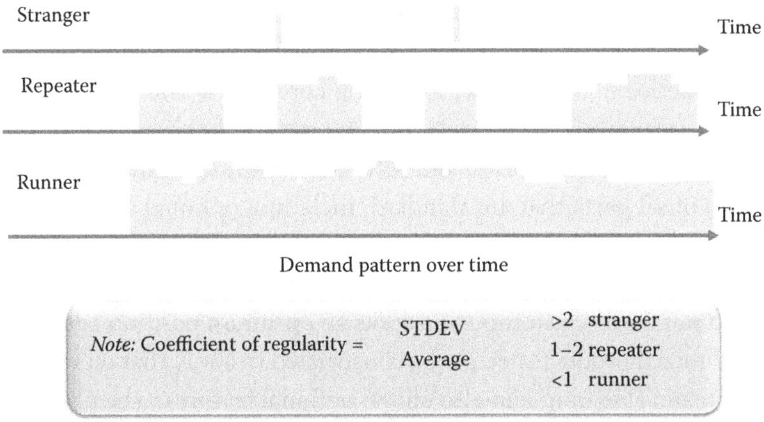

Figure 3.8 Demand regularity: runners, repeaters, and strangers.

for repeaters, a forecast and plan is likely to be best; and strangers need to be stocked if critical, but otherwise everyone needs to be aware of the lead time.

Recognize the demand regularity and use the appropriate techniques for runners, repeaters, and strangers.

* * *

3.5 Standardization

Standard parts are more easily obtained, usually cheaper, less likely to be obsoleted, and often more reliable. Standardization increases the volume we use and reduces the range. Where we can use common parts or raw materials it can dramatically simplify the supply chain and reduce costs. This is not only a matter of the cost of the parts themselves; it represents a dramatic reduction in the cost of managing the supply chain (purchase orders [POs], invoices, receipts, and inventory movements) that in many cases represents a large indirect cost. This is true of both the manufacturing and the service supply chains.

In one respect, this issue is a matter of providing design engineers with tools to help them find existing parts that are similar to the ones they need. Most modern computer-aided engineering (CAE) systems provide this type of facility. However, simpler approaches can also be highly effective (e.g., a spreadsheet of sizes showing which are manufactured standards and which are already in use on existing products). It is common *good practice* to use a *preferred fasteners* list, which is typically a list of standard sizes and threads and also indicates those which we already use. The same approach needs to be taken with all similar standard materials and components, from steel bar stock to resistors and capacitors.

Special parts are often part of our service differentiation. Sometimes uniqueness is necessary and the customer is prepared to pay the premium price. However, *customization* of an otherwise standard offering may deliver the same benefits from only a few special parts, while most parts remain standard. There are a lot of well practiced techniques involved in customization. Customization at the latest stage of manufacture is widely practiced, for example, for laptop computers. The main electronic circuit board, plastic casing parts, screen, keyboard, and many other items are all standard. The memory options, drives, accessories, and software usually contain the optional features. The proportion of all parts that are standard, including optional assemblies such as the electronic board, is very high. Laptops are configured at the final assembly stage and sometimes these are made to specific customer order. Figure 3.9 shows a continuum from unique to standard. Customization seeks an optimum position between them. The BOM for customized products need to be constructed in a way that represents the physical supply chain and assembly, and also allows optional features to be planned separately. This often uses modular bills based on a skeleton standard product, with additional bills for each option. These are described in the Logistics Section 5.3.5 Optional Features.

Customization also applies to the manufacture of metal parts. A basic form of standardization is the use of standard alloys and sizes. Beyond that, some parts can be produced from the same basic machining, and customized at a later stage. The advantages of this are reduced design time, reduced tooling and programing costs, and increased batch sizes through the first stages of machining and supply chain economies. Automotive companies are exploiting an advanced form of standardization with car floor designs. Sometimes, although they are a common design, they are not all produced in the same location and with the same tooling. However, the savings in component and tool design are worthwhile. Risks are also reduced because if a disaster occurred to one production facility, another one may be able to assist with the supply of common parts.

Design and manufacturing businesses should all have an approach to standardization. The work on developing and managing standardization is shared between different functions. Design, marketing, manufacturing, and supply chain all have a part to

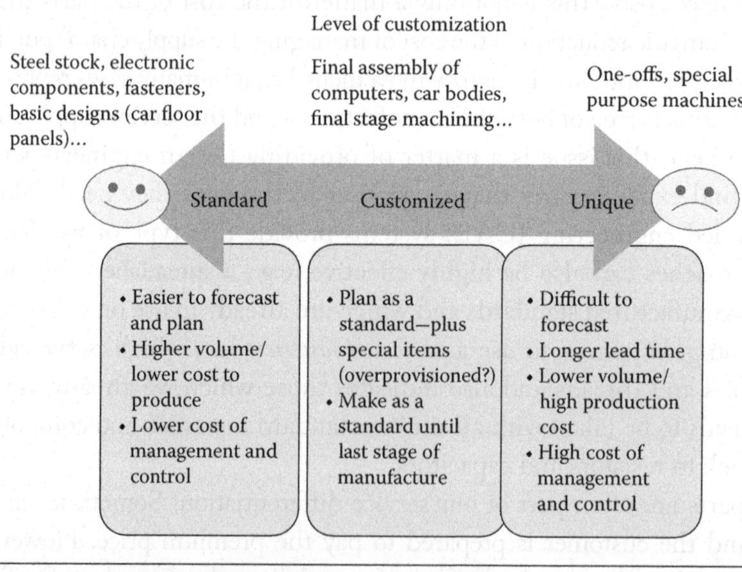

Figure 3.9 Demand characteristics: Standardization.

play. Design is probably the best equipped to implement, manage, and use standards. Marketing can judge where standard parts are acceptable and where unique or customized features will be valuable in the market. Manufacturing will be able to tell what the costs and other implications of standardization are (lead time, quality, etc.). The supply chain will be able to assess the implications on cost, quality, lead time, and availability from suppliers. I hesitate to suggest a standards committee, but a team with the seniority and practical knowledge and capacity to get things done is needed. This might seem like a costly overhead, particularly because it needs to include innovative, senior people who are valuable to the business. However, if you try to estimate the value this could deliver, it could easily be justified:

- Potential reduction in design and detailing cost due to reuse of existing designs and data
- Potential to get products to market earlier by using parts of existing designs
- Potential reduction in supply chain work due to reduced sourcing and purchasing work
- Potential saving in new tooling
- Reduced learning curve due to reuse of previous designs and parts

Multiproduct platforms are now a common way of reusing design, part detailing, tooling, manufacturing, and supplier development effort. The most obvious examples of this are in the car industry, where not only multiple models within a range, but also differently branded products, use a large proportion of the same parts. This is not only a way to get design engineers to reuse parts, it includes a range of large outsourced assemblies. This includes large areas of electronics, control systems, instrumentation, and interior fittings. The need to include suppliers in design teams and the implications for supplier selection are covered in Chapter 10 Product Design and Development and the Supply Chain.

The level of standardization, the approach taken, and the management of implementation and review should be by a senior and innovative board within all design and manufacturing businesses.

* * *

3.6 Critical Parts

Part criticality is used in many industries to assist with MRO provision. Critical parts need to be available immediately from inventory, but noncritical parts are likely to be acceptable on a longer lead-time.

The airline industry, for example, uses this approach in conjunction with failure mode and effect analysis (FMEA) to plan placement of spares on a global basis. Service agreements are typically made with maintenance organizations who hold spares on behalf of airlines using a range of client-owned and supplier-owned arrangements.

The chart (Figure 3.10) shows a simple approach used in support of a large operating plant. In this case, health, safety, and environmental considerations make it essential to keep certain plant functions working—so much so that some electric motors are backed up by *hot spares* that are installed and run alongside the working motors. Critical spares must be held by this organization, but there are over 100,000 potential spare part codes. Figure 3.10 summarizes a process for deciding which parts are critical. Implementation of this approach, principally the identification of critical parts, took nearly 12 months. Engineers are central to this criticality decision and a process was designed to use their time most effectively. Parts were added to spreadsheets and columns added for including a wide range of information known about the part, such as

- Where it is physically used within the plant
- Which control circuit it is part of
- What the function is of the part—operation, environment, safety
- Whether the functionality of the part is duplicated
- What the part type code is
- Who the supplier is
- Whether an alternative part can be used

Two-hour meetings were held where engineers reviewed the spreadsheets (projected onto the wall), and in conjunction with failure mode analysis took the decision to categorize parts as being critical or noncritical. Approximately 12% of parts were categorized as critical; a similar result has been achieved on a number of similar assignments.

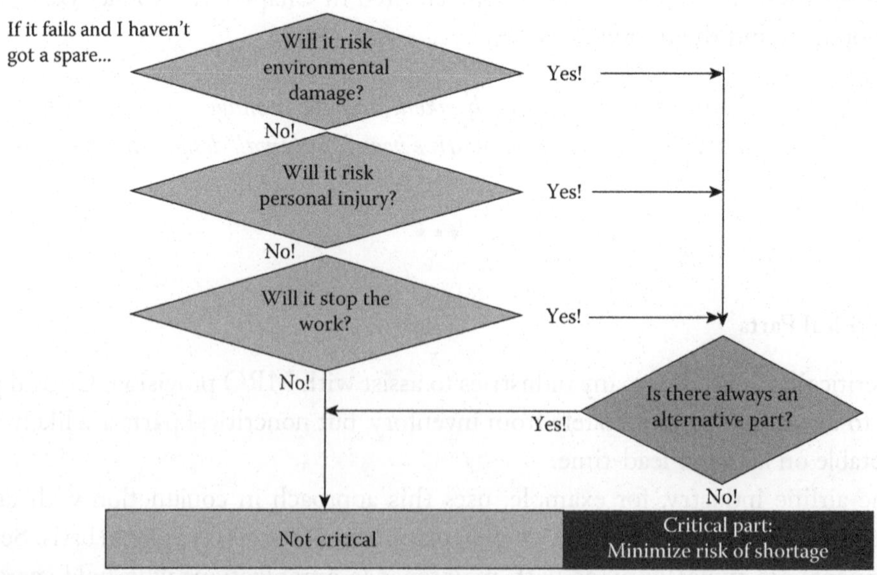

Figure 3.10 Criticality of spare parts.

For important plant and equipment, an analysis and listing of critical parts should be part of the design engineering and maintenance function prior to release of the product for shipping.

If you have a spares requirement for stranger parts on long lead times, assessing part criticality may be the only practical approach to inventory holding policy.

* * *

3.7 Forecasting and Planning for Different Types of Demand

Forecasting and planning processes are quite often treated like sausage machines. All products are forecast. All products are planned down to component level by MRP. This can be a waste of time—for example, if the supply lead time is shorter than the customer lead time, we do not need to plan it. We can organize supply when we have a customer order.

Figure 3.11 suggests some factors that can be taken into account when deciding how to approach forecasting and planning for different types of demand. We can use this to decide what inventory management approach to take on our range of products and parts. The right-hand column suggests potential approaches for each combination of characteristics.

The demand type example in line 1 is one we often see managed using MRP. The part is *critical* and the lead time is medium, so running MRP against a forecast should ensure we can meet customer demand. Typically the order intake varies from day to day and MRP is constantly adjusting the schedule to match real demand. The MRP calculations can be correct, but constant changes in the details of short-term demand

	Volume	Value	Regularity	Lead time	Dependency	Criticality	Potential inventory management solution.
1	High	High	Runner	Medium	Product BOM	Critical	MRP schedule for supplier planning. Kanban controlled delivery.
2	High	Low	Runner	Short	Product BOM	Non	Set-up flow of product, overseen by a VMI arrangement.
3	High	Low	Repeater	Long	Product variant	Critical	FORECAST to supplier for availability and schedule delivery to demand.
4	Low	High	Stranger	Long		Critical	Essential to project to order and schedule these parts early.
5	Low	Medium	Stranger	Long	Product variant	Critical	Ensure product option has long lead time + hold safety stock + expedite.
	?	?	?	?	?	?	?

Figure 3.11 Forecasting and planning for different demand characteristics.

can cause immense scheduling workload. This is also generally a waste of time. We can adjust MRP parameters and reduce the sensitivity to demand change. This will probably cause the schedules to be changed less often. However, MRP is simply not the right tool to use for short-term adjustments like this. A simpler solution is to use the MRP output to help the supplier plan forward, but to get him or her to control their deliveries using some form of kanban solution.

For the example in line 2, MRP is an even greater waste of time. The value of the part is low and usage is regular. The inventory risk is therefore very low. The lead time is also short. This is an ideal candidate for the vendor to manage using a shop floor bin system—one where he or she tops up the inventory every two or three days. The vendor does need to know the ongoing rate of demand (the *flow*). We can provide him or her with *gross requirement* data from MRP. However, unless demand changes significantly, the vendor can continue flowing his or her product and controlling deliveries using a suitable VMI technique. If the part is small, *bins* are often placed near the *point of use* and the supplier needs to ensure there is sufficient inventory to meet demand. If the part is larger, we may need to design some form of rack or special container.

The example in line 3 is more challenging. The part is an optional feature on the product BOM and is only a repeater. The lead time is also long and if we need to offer a short lead time to the customer, we need to forecast and schedule carefully. MRP can be used to provide a schedule based on forecast demand, plus some safety stock, but we do not really need it in our own inventory. If possible, a good solution would be to set up an arrangement with the supplier to make the product available to forecast, but only deliver when we have a customer order. To achieve this, we are likely to have to make special commercial arrangements with the supplier. For example, we are likely to have to *commit* to purchase all of the inventory the supplier makes for us against our schedule—even if we don't use it (a form of take or pay agreement). If it is of higher value, it may be a good subject for discussion at the sales and operations planning meeting.

The example in line 4 is a project-related part—perhaps a proprietary equipment of some sort. It is obviously essential to recognize this type of part early on in the project and make sure it is fully specified and procured in time. If the part is of high value we should not procure it too early in the project, but equally, we need to recognize the lead time and include its procurement in the project plan. Normal planning activity dates can be attributed (e.g., latest start and finish dates) and these can be modified as the project proceeds.

The line 5 example is an easy part to overlook. The long lead time demands early attention, but it is a stranger—likely to be a customer-specified option. The volume and cost are low and this makes it likely to escape attention. However, since the cost is low it may be acceptable to hold some stock. Otherwise it is important to make sure customer lead time expectations are managed. It may be a good idea to group parts with these characteristics under special planned codes (MRP output is grouped under these codes). Planners should take a careful look at these parts on a frequent basis to make sure demand is not outstripping supply.

The logic of this table is easy to follow, and despite its slightly complex appearance, it does show how demand characteristics can be grouped to help the selection of the best supply management techniques. You should make up your own table to suit your own requirements.

This approach and thinking has been an important part of quite a large number of inventory reduction programs with consultancy clients. The key issue is to have a clear and appropriate way of managing each set of demand and part characteristics. It is then essential to manage the selected way of managing the inventory well. Typically, this means managing carefully all the quantities for purchasing, scheduling, delivering, and holding in inventory. You could call this having a plan for every part (every part, every interval [EPEI]) rather than using, as the saying goes, a sledge hammer to crack a walnut.

Different part and demand characteristics demand different approaches to forecasting and planning. One size does not fit all.

* * *

4
Supply Management

4.1 Supply Base Alignment

The four-box supply base strategy shown in Figure 4.1 is similar to the Kraljic* model, the most useful way I know to start thinking about different types of suppliers. There are several similar diagrams (watch out for the axes being switched) using similar names to describe similar approaches. This one uses names (leverage, partnership, secure supply, one-stop) that describe the approach to working with the suppliers in each quadrant.

The horizontal axis is about spend with the supplier. Leverage and partnership are obviously areas of high spend. Secure and one-stop are low spend. I generally see partnership and leverage segments together representing about 80% of the total spend.

The vertical axis is more difficult to define. The complexity or difficulty of supply generally comes from technical issues. For secure and partnership suppliers there are likely to be only a few (say three to four) current and plausible alternative suppliers. A high score on the difficulty axis could also come from a shortage of a commodity or perhaps capacity of a particular type.

Positioning suppliers in the Kraljic box should be fairly easy using this sort of logic. Complex formulas for calculating the position of suppliers tends to move them all toward the center and talking about small changes of position within a quadrant assumes a level of accuracy that the process does not support. Changing technology and other circumstances also mean that some suppliers move from one segment to another. In my opinion, the analysis is very useful for assessing and developing the strategic approach to suppliers. In many cases, we can do things to move suppliers to another quadrant. However, discussing minor changes in position is not likely to be fruitful.

Partnership: High-spend items that are difficult to obtain are precious, particularly since they are usually important to the functioning of our product or service. Many businesses rely on these suppliers for their technical leadership. We need to get new technology very early in the new product development process. Furthermore, when supplies are short, we must ensure that we—not our competitors—get the product we need. We need the best efforts from partnership suppliers to make improvements in cost, quality, and delivery for our most expensive and critical supplies. These supplies are often crucial

* The Kraljic portfolio purchasing model exists in various forms but was described by Peter Kraljic and appeared in the *Harvard Business Review* in 1983. *Purchasing Must Become Supply Management*, Harvard Business Review, Sept. 1983.

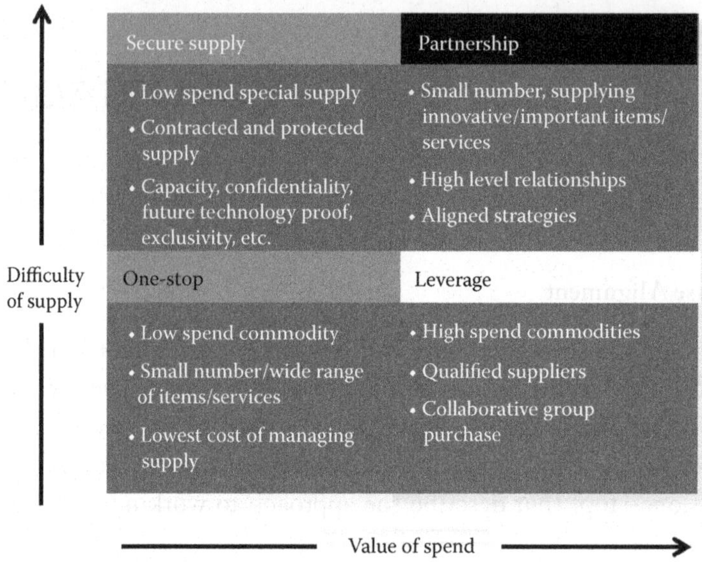

Figure 4.1 The four-box supply base strategy.

to our own business performance. Some form of collaboration on technical development, capacity scheduling, and performance improvement is often worthwhile.

Partnerships take time, effort, and a degree of trust, but the term *partner* should not be interpreted as meaning weak in any way. These relationships often involve more hard and straight talking than any others. Agreement is reached, sometimes in difficult circumstances, because of the relationship and some form of *shared destiny*. Relationships with partner suppliers need to include senior people from both supplier and customer organizations. They need to understand the critical nature of the supplier and be part of the relationship. There will come a time when they need to approve or in some way be part of a difficult decision on cost or commitment that is critical to the relationship. Mistakes at this stage that occur because a senior person does not fully understand the supply situation can be damaging. There is a limit to the number of partnership suppliers we can manage or need. These are often limited to about 5% of suppliers, but perhaps around 40%–50% of our spend.

Leverage: High-spend supplies that are easy to obtain (commodity items) are where we need to get the very best price. The quality and service level for these items *should be a given* but can be an issue for leverage along with price. We need to put suppliers into competition, consolidate our requirements, and use our buying power (high spend) to get the best possible price and conditions. High-spend, standard chemicals are a good example of a leverage purchase. When the specification is quoted, everyone knows the exact specification and standard of product required. The supplier does need to be qualified, but a quotation process is often what we need to get the best deal in this quadrant. The number of suppliers we use at any time needs to be limited, otherwise we dilute our volume to each supplier, and therefore our leverage. Having a larger number to consider when discussing quotations is valuable, but order volumes need to be shared between a smaller number to protect our volume leverage. The strategy needs to be decided carefully, but a good solution is often to select one supplier for most of the volume and perhaps two others for lower volumes in order to reduce risks and promote competition. Having a lot of suppliers actually supplying is not usually useful and tends to produce higher pricing and costs. In this quadrant, market conditions usually dominate pricing. A good understanding of potential market trends is needed, which may lead to some strategic planning of purchases—long order coverage or contract pricing where possible if prices are likely to rise, and short coverage and play the market if prices are falling. Our task is to leverage suppliers to get the very best deal we can. Be careful though: when the market changes and supplies are short, suppliers in this quadrant do enjoy *payback time*, sometimes with entire justification if we have behaved badly.

Secure supply: The most difficult quadrant is where we have low spend on items or services that are difficult to get or may even be from a *sole source*. These are often technically challenging items or services. They may be critical to the business and often demand a lot of effort for a relatively low spend. We certainly cannot do without these suppliers. They are often small and can be vulnerable to bankruptcy or takeover by our competitors. The general supply situation is likely to be a high risk. Sometimes we need to challenge why such suppliers have ended up in such a critical and commanding position. For example, sometimes a subcontractor has made modifications to parts in order to make them work for us, but we have not properly recorded what has been done. Over a number of years, the supplier can become the only one who knows exactly how to make or do something for us. The process may be repeatable by other suppliers, and the purchase should perhaps be a straightforward *leverage*. However, the exact requirement or specification is not known and we are therefore trapped into a sole source. These situations need to be challenged. True *secure* suppliers do have something unusual and that may be critical to our business. Our task is to secure the supplier, particularly where they may be vulnerable to failure or takeover. We need to talk to them regularly, understand the vulnerabilities of their business, and manage the risks. We should also be aware that we may be the prime cause of some of their difficulties through poor scheduling or late payment. If all else fails, we may sometimes need to buy a supplier's business in order to secure supply; for example, if he or she is in danger of bankruptcy. Alternatively, we may enable another supplier to purchase the business.

One-stop: Easy-to-obtain, low-spend items are still vital to the success of our business but need to be organized to require low administrative effort. There are usually a large number of purchases in this category—more than in any other quadrant. This represents a large amount of work, but for only a small proportion of the purchase value. Supply arrangements must therefore be simple, organize excellent availability, but require little or no management effort. These items are of low price. Price reduction will deliver little benefit, but cost reduction (usually the cost of administration) can be substantial. We have all seen examples of the 50-cent item for which it costs a total of $70 to raise the PO, receive, process, and pay the invoice. Madness! Low-value transactions can be managed effectively in a number of ways, providing excellent service levels and costs, and sometimes freeing supply chain people from almost pointless bureaucracy to do more valuable work. Alternative approaches for managing low-cost items are described in more detail in Section 4.2 and also in Section 5.4.1.

Use the four-box strategy (Kraljic type) analysis to categorize your suppliers as a basis for supplier management and relationship development.

* * *

4.2 Managing Low-Value Items and Services

Low-value parts and services represent a disproportionally high part of our bureaucratic workload. This can be a major cost. Figure 4.2 shows how transaction costs can be disproportionately high for lower cost parts. Service levels in the supply chain, warehousing, and accounts payable can be degraded because of the large number of these transactions.

Alternative supply chain processes must deliver these purchases with much lower administrative burden and yet provide high levels of financial control, quality, and service.

Low-value purchases differ widely in type and importance. Having them available may be just as important as having high-value parts. A high-value PCB cannot be fitted without the low-priced clips that hold it in place!

When it comes to how to manage low-value parts, *one size does not fit all.* For example, low-value parts that are used regularly can be managed by some form of vendor managed inventory. Irregular requirements could be purchased using a P-card. Here are a number of solutions to managing low-value procurement, each of which is suitable for different types of *buy.*

4.2.1 One-Stop Shop

A large proportion of low-cost purchases are frequent and of commonly available commodities. Under the *one-stop shop* arrangement, a supplier maintains his or her

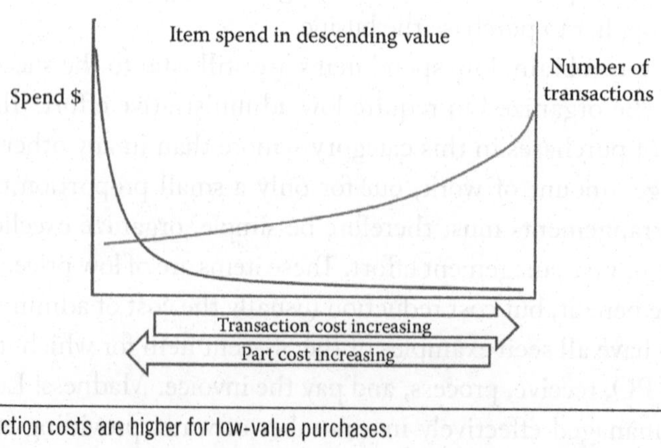

Figure 4.2 Transaction costs are higher for low-value purchases.

own stock in a shop area that exists within the customer sites. The supplier issues to employees at a counter against appropriately authorized requisitions and records them on his or her system. At month's end, a single invoice is issued, accompanied by data (typically a spreadsheet) showing all issues made within the month, the items issued, price, customer, approver, cost center, date, and time. This enables the finance department to make appropriate cross charges and pay the invoice. The supplier is unlikely to make additional charges for this service if the sales revenue from the one-stop shop is reasonable. Since the supplier's space, heat, and light are provided free of charge, favorable prices are also more likely. However, total cost, not price, is the real issue and arrangements should provide the lowest cost and highest service level for customers. We have seen that suppliers running a one-stop shop are incentivized to maximize service level and increase the range of goods and services they offer—it increases their sales! This arrangement can be helpful in factories, engineering and maintenance sites, hospitals, large office blocks, and all other areas where regular items are required on an ad hoc basis (e.g., they are not a piece-part of a manufactured product).

4.2.2 P-Cards

P-Cards are a very convenient way of paying for goods and services. They are similar to a personal credit card. A business can set up a P-card account with a number of cards, each allocated to a person within the organization. This person may then purchase parts and services in a similar way to a personal credit card. However, as you would expect, there are far more security measures available to restrict the purchases. For example, there are credit limits, individual transaction limits, and limits for spend within a particular period. In addition, there are *merchant codes* set up for each card that determine the types of supplier (*merchant*) with which the card can be used. Figure 4.3 provides some examples.

The last example in the figure, "5122 Drugs, Drug Proprietors, and Druggist's Sundries" is an example of a merchant code that a company has decided will not be allowed on any of its P-cards. When cards are issued, they are set up with the *merchant codes* that match the purchases we require this person to make. Failure to do

1711	Air conditioning contractors—sales and installation
1711	Heating contractors—sales, service, installation
1731	Electrical contractors
4900	Electric, gas, sanitary, and water utilities
5013	Motor vehicle supplies and new parts
5021	Office and commercial furniture
5039	Construction materials, not elsewhere classified
5111	Stationery, office supplies, printing, and writing paper
5122	Drugs, drug proprietors, and druggist's sundries

Figure 4.3 Example merchant codes.

this properly may enable the card to be misused. We believe the supply chain should be responsible for organizing this process in cooperation with finance. P-cards carry a good level of security if they are set up properly. However, the supply chain should be responsible for monitoring spend and checking for misuse of the card.

The P-card provider is likely to pay the supplier almost immediately for goods or services supplied against his or her cards. This is a great benefit to suppliers who are short of cash or whose cash flow is tight. Of course the card provider (usually a bank) charges the supplier a percentage for this service.

The bank usually requires payment from the business holding the cards at month's end. Full details of all purchases are provided, usually electronically. The business can then check the purchases and make the appropriate cross charges against departmental budgets.

Who should be provided with a P-card? Some businesses only allow them for their purchasing staff. However, this means that requirements first have to be communicated to the purchasing department and will inevitably incur a time delay. Any person who needs to purchase items on an irregular basis and who has a cost center should really be responsible enough to hold a P-card.

We have found financial controllers reticent to allow P-cards at first. They are mostly concerned that control of purchases will be lost. However, we have been able to convince them of their benefits and reassure them about security by explaining how the cards work, especially with merchant codes.

4.2.3 Buy Houses

Buy houses are arrangements that effectively *outsource* the purchasing function. Buy houses will make purchases for a business and apply a markup. Buy houses are not an effective solution for regular purchases that can be supplied by vendor managed inventory (VMI), or where the user knows the likely sources and can hold a P-card. In some poor examples of implementation, buy house implementations almost double the purchasing administrative burden (although the number of invoices can be reduced). However, they can be very effective when the suppliers for parts or services are not known. This is particularly true with specialist buy houses. An example from a client assignment was from a land-based supply chain person supporting marine vessels. He had a request for a replacement bearing for a deck winch. He was told the manufacturers name, but no other details. He contacted a buy house specializing in marine spares who asked for details from a plate that they said would be beneath the rear of the winch. The crew checked the serial number and other details from the winch and relayed it back. The buy house then told our supply chain person that the winch design had been updated and a new end plate, seals, and fixings would be required. These were successfully purchased through the buy house and a markup of 6% was charged—saving the supply chain valuable time and getting the supplies delivered quickly. The only issue we would have with this example is that we prefer to use a *flat fee* with buy houses; they are not encouraged to give us the best prices by paying them a percentage!

4.2.4 VMI

VMI is an arrangement that is often used in manufacturing facilities and retail businesses. A range of techniques can be used to signal the need for stock replenishment to the supplier. On one of our assignments, the supplier manufactured a special packaging component for the maker of a famous food product (see Section 5.4.6). In this example, a web based system relayed inventory information from the customer's system to the supplier. However, many VMI arrangements are much simpler arrangements, sometimes relying on kanban. My colleague Crispin Brown quotes an example which uses a web cam permanently looking at the inventory and relaying the picture back to the supplier!

VMI can equally be used for *indirect* purchases. For example, an inventory of frequent workshop supplies such as *releasing fluid* and *Loctite* can be owned and maintained twice weekly by the supplier in the workshop. If required, this can be under lock and key held by the workshop manager, but is available directly as required. Monthly invoicing can be checked using a simple system. VMI covers a wide range of techniques for vendors to manage inventory for their customers. Parts with higher value can also be managed with VMI.

4.2.5 Vending Machines

Vending machines are included in this section because they reduce administrative costs and are suitable for a range of semiregular items. They are quite secure, and therefore suitable for slightly higher value items. Vending machines can be used for BOM and non-BOM requirements, although they tend to be for ad hoc requirements. Some typical examples are

- On the shop floor for glue for assembly operations, machine tool inserts in the machine shop, and medium-value disposable personal protective equipment
- In offices for toner cartridges and higher-value stationery items
- In hospitals for syringes and other higher-value or secure items for nursing staff
- In public spaces such as airports for selling medium-value consumer goods (e.g., iPods)

See Section 5.4.5 for further discussion.

4.2.6 Risks to Critical Supplies and Suppliers

Mixed with the large number of low-spend commodity suppliers are some which, although low spend, provide services or parts that are critical to operations. There are dangers if the management of these suppliers is outsourced, say, to a buy house. During implementation of new low-spend systems it is important that they are identified and set aside for appropriate special management. Examples of these are typically suppliers who supply special technical parts or services. These often require discussion of specifications or requirements and are unlikely to be properly understood by a buy house unless they are a specialist.

Avoid spending all your time and resources on administration of low-value orders—use one of the alternative processes described here.

* * *

4.3 Categorizing Spend and Starting Work on Your Supply Base

It is important to categorize commodities and suppliers as the basis of supply base development. It is a key supply chain's responsibility to optimize the supply base to meet business requirements. Historically, most businesses have had too many suppliers. The task has been *supply base reduction.* The real task is more demanding. The supply base profile needs to match business needs for technology, commodities, and specialist services at a cost-effective and suitable service level.

A supplier spend report is one of the best places to start. Like all data, this may require some *cleaning up.* Simple things like multiple entries for the same supplier (spelled differently, even with or without capitals) need to be corrected. If we are using data from the purchase ledger, other spends such as tax and employee expenses may need to be stripped out as well.

Figure 4.4 describes how to differentiate the types of supplier and what action then needs to be taken with each group. It is evident that actions are different for each type of spend.

Figure 4.4 shows suppliers organized by descending value of spend. Obviously, partnership and leverage spends are at the high-value part of the curve. Secure and one-stop suppliers are at the lower-value end.

The recommended course of action for each is described below.

We need to sound a special word of caution about secure suppliers. They can be positioned anywhere along the lower part of the spend curve and are sometimes difficult to spot. They are also a high-risk group and therefore much care needs to be

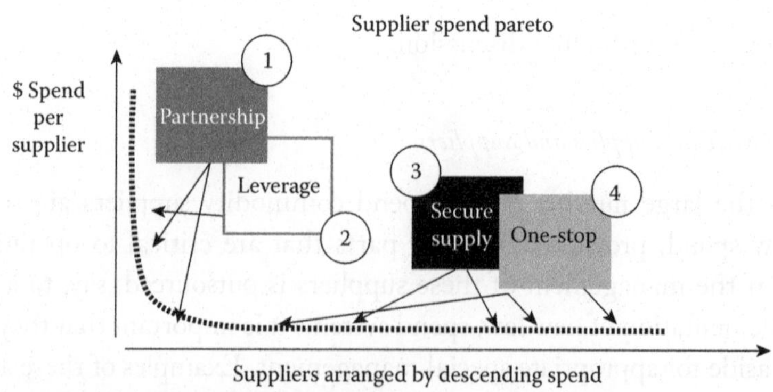

Figure 4.4 An approach to aligning your supply base.

taken to make sure they do not get included with one-stop suppliers, and they must be dealt with accordingly.

For partnership suppliers, engage the likely stakeholders in engineering, manufacturing, and marketing, and check that you have clearly understood why they should be a partner. They may not really warrant this status, but may have engineered themselves into this position by adding special features to the product for which only they know the details. Remember that a leverage supplier will try hard to become a partner because it offers at least a little shelter from competitive leverage. Once established as a true partner, a supplier needs to be managed appropriately. We need to clarify the following:

- Who will manage them?
- Who are the important stakeholders?
- What is the technology roadmap?
- What is the basis of cost/price?
- How much of their capacity do we need and how much do they have?
- What are the supply risks?
- What will we need from them and what will the supplier need from us?

If the supplier is a sole source, we need to mitigate against risks such as a disaster in the supplier's own business, or on a lesser scale, the possibility of a shortage of capacity. These suppliers are likely to be important to our business success and they need to be managed actively and with a clear agenda or plan. Partnership suppliers all warrant a development plan and most will really benefit from some form of joint working arrangement and commitment. We should be careful to target the development plan on the real needs of our business. It is really unfortunate to have worked closely with the supplier to achieve excellent cost reductions only to find that the marketplace is looking for additional features rather than lower cost!

For leverage suppliers we should make sure we can obtain the capacity we need and apply appropriate leverage to get the best possible prices. We have to think about how many suppliers we need for each type of commodity. We may need to consolidate the supply base down to using only one, two, or three suppliers in order to achieve the best volumes for leveraged pricing. Our strategy may be something like 80% of the business to one supplier and 15% and 5% to the other two in order to promote the leverage. It may be useful to keep in good contact with several more for use at the next round of leveraged sourcing. Changing supplier from time to time may also be a useful strategy because those asked to quote successively without ever winning any business will soon lose interest. I often found a change in leveraged suppliers beneficial from time to time anyway. Managing leverage purchases can be brutal. However, changing supplier for a 1% lower price can be worth the disruption if it will deliver improved profits (remember how much impact large spends can have from Section 1.1.1). This does not mean we should ever be less than completely courteous and behave with integrity to all suppliers. Apart from my belief that this is the way to behave, most poor behavior will not count in your favor in the end. We also need to remember that market conditions may change. Shortages of capacity for example may turn the leverage around the other way, which may result in the start of the get-your-own-back season for suppliers. You have been brutal when negotiating prices, but with courtesy and integrity. Although you are likely to suffer higher prices in a subsequent shortage situation, at least you should get the product. Others may not.

It is very important to identify secure suppliers among the (usually numerous) one-stop/commodity spends. Their status does need to be challenged. As described in Section 4.1, they may have engineered themselves into this position by not telling us about all the small tweaks they made to the original design (sometimes our own). These situations are difficult to retrieve. They are likely to require engineering effort to at least specify what the part or process is before it can be re-sourced. This may not be worth the cost. However, unless or until the part or process does become commoditized it must continue to be treated as secure supply. True secure suppliers represent a risk to the business and it must be ensured that they are exactly that—secure businesses. If possible, personal visits on a regular and informal basis can be a good way to check the supplier's business. I find it easier to discuss the supplier's general business level, capacity issues, cash flow, and so forth on this sort of basis. Visit him or her regularly to make sure that the business numbers are healthy and there are no other recognizable risks. In some cases, particularly if they are vulnerable, we may need to take action. We know several examples of work being given to suppliers just to keep them in business. We could even buy the business or we could arrange a takeover by another trusted supplier. This is a good example of how we may need to reengineer the supply base to reduce risk.

We use the terms one-stop to describe the large number of low-priced commodity buys we make because it would be ideal to source it all from one supplier. The cost of administering all the purchases is often more than the value of the items themselves. We described a number of different alternative processes for these spends in Section 4.2, including an actual one-stop shop on the customer's site. We need to examine the nature of the purchases and develop a range of appropriate supply arrangements. Figure 4.5 gives some examples of a simple way to start.

There are numerous ways the detailed processes can work—and it is a lot of fun working out how.

One key problem we are likely to have to solve is that of too many suppliers. This could just mean we need to go through a selection and deselection process. This is likely to work in a similar way to a normal sourcing exercise. If we need to keep a number of suppliers because we need their product, *tiered* or *clustered* supply arrangements, as shown in Figure 4.6, may be the answer. This works well for sets of parts

Use of item	Regularity	Potential solution
Direct supplies—for product manufacture	Regular	VMI solution such as plastic bins on shop floor OR include in a kit
Indirect supplies	Regular	One-stop shop
Direct or indirect	Irregular	P-Card

Figure 4.5 Examples of appropriate solutions for one-stop purchases.

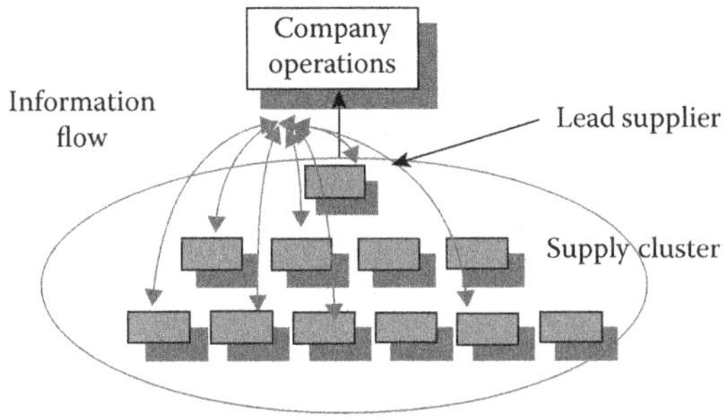

Figure 4.6 Tiered supply and supply clusters.

for direct manufacture. One of the suppliers acts as the lead supplier—organizing the schedules, collecting all the parts, collecting payments, and paying all the suppliers in the tier. There are a lot of ways that detailed arrangements can be made. For example, sets of parts can be delivered in kit boxes, racks, or even assembled.

The scheduling, supply, receipts, and payments are all managed by the lead supplier. The lead supplier manages and administers all the other suppliers in the group. We first heard of this concept being used in automotive manufacture and later as it was introduced into aerospace supply chains. This concept is typically arranged to supply one work area such as a subassembly stage. One early example was the outsourcing of an instrument panel assembly. The lead supplier was responsible for obtaining all the metal and plastic parts, switches, instruments, and wiring from the supply tier, assembling them and shipping them to line-side ready for use. The lead supplier was given the schedule information and was paid for the whole assembly. He then managed and paid all the component suppliers. This is clearly a way of reducing our administrative and management tasks by passing a lot of it to the lead supplier. This works very well in practice although lead suppliers need to be selected carefully. The management of our supply base is likely to present new challenges to them, including the need for new resources and costs. Figure 4.6 shows that lines of information flow down to most of the tiered suppliers. This highlights one of the issues first encountered in the instrument panel example above: contact with suppliers was reduced, including those who supplied the most technical parts (e.g., switches and instruments). The terminology was adjusted at this stage from tiered supply to supply cluster. This was done to recognize that contact with these suppliers needed to continue on technical, quality improvement, and similar issues.

Use your supplier spend report and the four-box strategy and determine which quadrant each supplier is in—then use an appropriate relationship for each supplier.

* * *

4.4 Commodity Management

Almost all supply chain organizations use some form of commodity or category management. The objective is to enable a senior supply chain person to be made responsible for the strategy and development of a particular type of purchase (e.g., electronic components, fuel, plastic moldings, machining, people, travel, wool, chemicals, facilities, and power). Others will then take responsibility for the more tactical work such as raising purchase orders and scheduling deliveries.

Figure 4.7 shows the typical interfaces between commodity management, sourcing, and supplier management. These are three distinct roles, although sometimes scale allows one person to do more than one of them.

The role of commodity management is to coordinate the strategic development of the supply base for a commodity group on a global basis. Responsibilities typically include

- Collecting overall demand and ensure availability of sufficient capacity
- Defining the commodity strategy relative to the four-box strategy (partner, leverage, secure, one-stop)
- Defining the target number of suppliers and type of relationship necessary
- Agreeing with stakeholders and defining the technology roadmap
- Agreeing with and defining any key technology issues for future development
- Agreeing to essential contract contents and issues to be covered
- Coordinating strategies for risk management, cost reduction, and performance improvement

These strategic decisions pave the way for sourcing of the most appropriate suppliers (covered next in Section 4.5). This description and Figure 4.7 should not give the impression that the three roles described are discrete and successive. There needs to be information flowing in both directions. In the process of searching for potential sources,

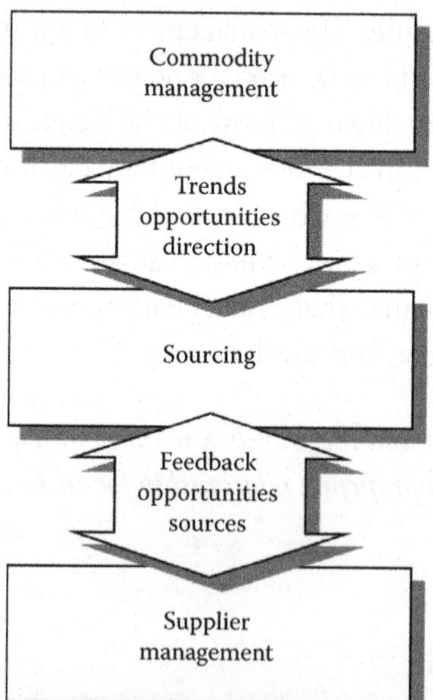

Figure 4.7 Commodity management.

new opportunities often come to light that are so significant as to change the category strategy. For this reason, there must be an excellent level of dialog across the three roles.

The supplier management role is to optimize the performance of suppliers. This includes the tactical issues to support solving problems with delivery schedules, quality issues, and so forth. However, on a more managerial basis the role should include

- Measuring the important aspects of supplier performance and driving performance improvement actions
- Driving and managing supplier development activities to meet current and future business requirements (e.g., from the engineering, manufacturing, quality, and marketing groups)
- Developing, agreeing on, and managing the supplier agenda or plan that optimizes future performance
- Providing data and qualitative feedback on trends and opportunities to commodity management

The nature and size of these roles will differ depending on the businesses size and scope, and the commodities. The amount of the work which is sensible to do locally or in a group headquarters will also differ.

For example, for a large, global business using a significant quantity of chemicals, which are a global commodity, it is likely that strategic sourcing, technology roadmap management, contracting, risk management, and even top-level capacity management can be done globally. These are the commodity management and sourcing roles. Supplier management at the local level then needs to take care of tactical issues and local supplier management.

For some large organizations travel is managed as a global commodity and users make their own bookings online, leaving very little for the local supply chain to do. Here again, a large amount of the work in all three roles can be centralized.

It is possible to source the provision of catering globally. Several catering and service suppliers have global capability and this has been quite successful for some large businesses. However, food is a very local issue and the detailed organization of menus and some aspects of service level must be left to local supply chain organizations.

A more classical application of these roles would be metal machining. For a large business, commodity management and sourcing roles need to be global. The supplier management role, however, needs to be done closer to the action where tactical issues can be seen and managed, and where there is sufficient understanding of ground-level issues to properly manage improvements. Even here, the level of feedback to the commodity management team needs to be substantial.

Another classical application is electronic board manufacture (PCBs). The commodity manager usually has a busy role in coordinating the developing technology roadmaps. New technology developments are quite rapid and may require re-sourcing. Implementation may be phased and old and new technology and suppliers may run in parallel. New design rules (see Section 4.10) will be developing, and therefore, communications between the suppliers and engineering and manufacturing need to be kept up to date and managed. Sourcing and supplier management are equally busy in this commodity area and need to be coordinated by the commodity manager.

Depending on the situation, the roles can therefore be covered by one person or by several people. Some of the roles can be centralized and some cannot. However, it is important that all three roles (commodity management, sourcing, and supplier management) are recognized as distinct and be done well. For example, the temptation for a commodity manager to get involved in tactical price negotiations with a local supplier will both undermine the local supplier manager and prevent the commodity management role from being done as well as it could be.

Group your purchases into commodities for which a commodity manager can be given strategic responsibility.

* * *

4.5 Sourcing and Supplier Selection

These days, sourcing usually means global sourcing, which brings with it a wide range of issues. Beyond the normal capability and suitability of the supplier, the costs of logistics, and the effect of lead times, considerations about political climate and stability, taxes, and trade agreements come into play. It is vital to weigh these well in the assessment because they can quickly tip what looks like a favorable source into a costly mistake.

Labor rates are used as one of the guides to where some of the lowest cost sources may be. General ratios between Central Europe, Eastern Europe, the United States, China, India, and other areas are useful where direct labor is a key part of cost. Furthermore, the cost of indirect labor such as engineers, managers, and administrators, and even other indirect costs such as rent and a wide range of services tend to be lower.

Even if you are a real *trail blazer* and you intend to get competitive advantage by using a new area before everyone else, the availability of materials, appropriately skilled labor, and infrastructure need to be considered carefully.

A client aiming to do this selected China as the key new area to source some complex machined parts, moving a fair proportion of manufacturing from Europe and the United States. There were good machining companies available and some were selected to start work. Projected savings were substantial. However, some of the steel alloys required were not available from Chinese mills and needed to be imported. This extended the total lead time, and along with the 5 weeks of shipping time for the finished product, increased the reliance on forecasting and scheduling accuracy. Our client then found Chinese mills who made the required alloys. However, although the percentage alloy contents for the Chinese steel was the same, the tolerance on the alloy contents were not. The slight differences in the percentage tolerance on alloy content required components to be requalified. Engineering time to validate the alloys was expensive and could have been used more fruitfully. Our client is also having difficulty obtaining design prototype machined parts close to the design teams in the United States and Europe that are then transferred to China for volume manufacture. Suppliers are able to provide valuable input to design engineers, enabling components to be made more easily and cheaply. However, this input needs to be provided by the suppliers who will manufacture the volume product—not the suppliers who only make the prototypes. Our client's search for an ideal solution goes on. A local European or U.S. machinist with volume production facilities in China, or a partnership with a Chinese machinist, could be a practical solution.

This is typical of the difficulties of global sourcing. It indicates the level of thinking about total costs, lead times, and product design needed to truly evaluate sources.

Regional costs change over time. A long time ago, we used to buy computer monitors from one of Japan's largest companies that were manufactured in Nagasaki. It became too expensive to manufacture in Japan and the factory was moved to Singapore. Singapore is an attractive place with good infrastructure, logistics connections, and a skilled workforce. However, this rapidly became too expensive and the facility was moved again, this time to Thailand. The need to move from Singapore occurred after only 3 years. This may mean Singapore was a poor choice. However, labor rates, capacity, technical capability, and political changes continue at an increasing rate. This means that the supply chain people need to look ahead as far as possible when selecting regions for sourcing. It also means we need to factor in the associated effort and costs that re-sourcing incurs.

Another example of the complexities involved in making bespoke parts abroad is from a client who outsourced manufacturing to a range of suppliers in the Far East, particularly in Thailand. The suppliers work hard on product quality, schedule changes, and adherence and costs. A new range of high-quality plastic molded components were required. Local molding suppliers were available, but the molder relied on tool-making facilities he used in Taipei because he could not find the right quality of toolmaker locally.

Sourcing more standard items rather than bespoke on a global basis is of course simpler. In many cases the example of total costs given later in this chapter represents a large part of the considerations. The example assesses the majority of the costs comparing European manufacture with the Far East. In addition, some other suggested items to be considered are

- Are the levels of cost likely to change? (China has a huge population and the inertia required to change costs is significant; however, Singapore has a limited workforce and a client has recently needed to recruit labor in China to work in Singapore)
- Are the conditions on tax and other trade regulations likely to change?
- Will the suppliers in the region keep up to date with technology development in the commodity (and how important is this to your business)?
- Are there risks to supply from political, trade restrictions, environmental, or commercial issues?

Sourcing and selection of suppliers requires assessment of a wide range of issues. This is certainly not just about price, although final evaluation may be expressed in terms of total cost. The stakeholders whose needs must be taken into account range from marketing through design and engineering, environmental health and safety, manufacturing, and logistics. Some stakeholder representatives may need to be part of the evaluation team for the more important considerations.

Many large companies have a web-based system to enable potential suppliers to provide information. This is usually entered through some sort of supplier portal and an early requirement is for the potential supplier to sign a nondisclosure agreement (NDA). After this, standard information is collected from the potential supplier on the business and what they supply. The potential supplier can usually complete a free format entry where he or she can convince the buyer that they have something worth looking at. The involvement of supply chain people in this type of up-front process can make the process more effective and avoid introducing them into a situation late in the day and trying to get them up to speed quickly when many decisions have been semimade on shaky foundations. At this stage, good relationships and communication are needed between supply chain and other departments. Depending on the organization, this may include sales and marketing, operations, or in fact almost any function.

It is a good idea to get together a first view of the broad requirements at the data research stage. Along with this, the collection of information about potential suppliers in selected regions starts to provide a view of what is available. We have a range of Internet facilities available now that supplement trade magazines and directories. As this group starts to discuss all the possibilities it will enable feedback and refinement of requirements.

At this stage, we should be forming a fairly accurate idea of the type of supplier we are looking for:

- What size (ideally one that gives us enough *weight* but not be too much so that the supplier will be too *dependent* on us)
- What range of products, facilities, and expertise
- What capability for design and technical development
- What location(s) and capabilities for global service
- Someone to make one item or service and or someone to provide a significant part of a commodity group
- What levels of quality
- What levels of cost
- Machines/equipment/capabilities
- Other customers/references
- Technical approvals

The important stakeholders need to agree on the evaluation matrix, which is no more than a list of the decision criteria (usually in sections such as *technical, equipment facilities, and capacity*). Figure 4.8 summarizes the evaluation process, showing the range of requirements that may need to be taken into account, and an agreed

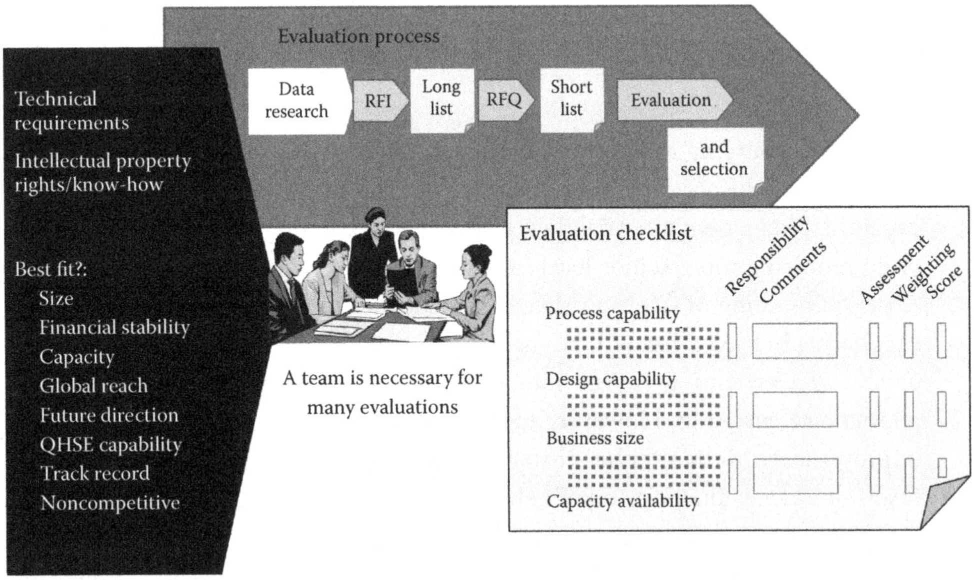

Figure 4.8 Supplier evaluation and selection.

evaluation matrix. Our evaluation matrix will become the criteria for supplier selection and we will need to agree on the relative importance of criteria (usually a weighting score). We should also establish who is to be responsible for leading the assessment of each of the criteria. In some countries, and in particular where the work is for government contracts, the selection criteria must be published at least to those participating in the tender (I think it is a good thing to tell the potential suppliers the selection criteria anyway). In some cases, suppliers have taken legal action against a decision not to select them and have queried the decision—and in some cases won compensation.

If we have established our selection criteria and a reasonable number of potential suppliers (e.g., 20 or 40), a request for information (RFI) may be a good next step. This will enable us to reduce the number of potential suppliers to a more manageable number for detailed analysis. If we have not already done so, we should ask them to sign an NDA or confidentiality agreement. The issue at this stage is unlikely to be intellectual property; however, a mere inquiry may reveal our intention to do something we do not want our competitors to know about—volumes, new markets/regions, or new products or processes. These can all be revealed through an inquiry. The RFI can ask for some general information about the business, including financial data. This is usually difficult to get from published data; for example, the data for a specific division we are interested in doing business with can be summarized up to a division or group level. We can then ask for information about the product or service we are interested in—is it the type we need, what are its features, how many do they sell (and to who), some outline pricing, and so forth.

We should work with the important stakeholders to analyze RFI responses and thin down the possible contenders into a *long list*, maybe up to about six in normal circumstances. It is often useful to have a conversation with the potential suppliers before issuing a request for quotation (RFQ). In general, the type of questions we need to start getting answers to will differ depending on the position of the supply in the four-box strategy:

- *Partnership* sourcing requires evaluation of longer-term issues such as strategic alignment of objectives, potential for competitive issues, technical and process capability, management capability, global reach, and quality of the business—often requiring more senior level involvement.
- *Leverage* sourcing of freely available commodities requires quality and technical checks, but selection will be mainly based on cost.
- *Secure supplies* tend to select themselves because there is little other supplier choice available. However, an evaluation process can sometimes lead to improvements in the supplier's capability *or* remove the need for secure supply.
- *One-stop shop* evaluations are likely to be more concerned with availability, global reach, and customer service capability. User involvement is often beneficial in helping with the evaluation *and* in gaining acceptance of changes in practice from users.

We need to test the potential supplier's competiveness. It is typical to use either the parts or services we want or a sample if the range is larger. To enable realistic prices to be given by the supplier we need to give an estimate of the total volumes we require. We need to be careful not to make this request appear in any way as a commitment to either purchase these volumes or even purchase at all. It is useful to obtain a structure or *basis of price* to underpin the quotation with information about how the cost is built up and importantly, how it may vary in the future if circumstances change. This is often useful for material content, fuel content, or in fact any commodity or service that is volatile in price or changes significantly with volume.

In addition to testing competitiveness, we also need to know what they are like as an organization. We may need to know what they will be like as a long-term partner, or we may just be concerned with a specific project. Either way, we can ask for other information about quality, technical investments, policies, ethics, and so forth.

At this stage we may be ready to issue a request for quotation (RFQ) or a request for proposal (RFP). An RFQ specifies our requirements and generally requires a bid which is compliant with our specification. An RFP defines a requirement or problem to potential suppliers and asks them for their proposed solution.

Before issuing the RFQ, your selection criteria need to be complete and agreed on with your stakeholders. In addition, you need to have agreed on their relative importance. Pricing (or better still, costs) will be important, but the technical suitability of their offer may be even more critical. Their capacity, quality approach, and results may reveal key issues about the way they run their facilities. If particular technologies are involved, for example electronic board assembly, you need to involve specialists to assess their capabilities. Your work may involve design, in which case get someone who knows about design, computer aided engineering (CAE) systems, and your own design standards on your team. Their suitability to meet your requirements also includes things like their ownership and the size of their business. There is a *sweet spot* for the proportion of a supplier's business that one customer represents. This varies with the type of supply, but generally, I am uncomfortable at much less than 5% or over about 20%. Otherwise, the supplier will be too dependent on us. If, for example, our business increases a lot, they are unlikely to be able to deliver the additional capacity. Figure 4.9 shows a range of inputs to be considered, the RFQ/request for proposal (RFP) process and the functions potentially required to support the process.

RFQs will solicit compliant quotations such as those which meet the specifications and conditions we define. Although the lowest cost or price will be important, we normally include a provision in the RFQ that we reserve the right not to choose the cheapest. There may be other issues we need to consider as part of our selection. Within the RFQ, however, we will have specified our requirements as closely as we can to ensure the prices we are quoted are for the same goods or services. We need to define our requirements for areas such as

- Specification of the goods or services
- Warranties required
- Timing of supply, our required schedules, and our flexibility to change
- Contractual terms and conditions

RFPs should solicit the supplier's best approach to solving your needs and these could vary a lot between suppliers. This allows potential suppliers more scope to be inventive than with a more tightly defined request for quotation. What we *cannot* then do is pick the best bits of the proposals and publish them in a requote to a range of suppliers. That is effectively stealing their ideas. It is illegal. You are likely to be sued and they are likely to win. If it would bring a lot of benefit and you really need to combine the ideas, you must organize for them to do it. For example, you could introduce two potential suppliers with the best ideas at a meeting and leave them to explore the possibilities themselves. We are likely to be best saying to both parties

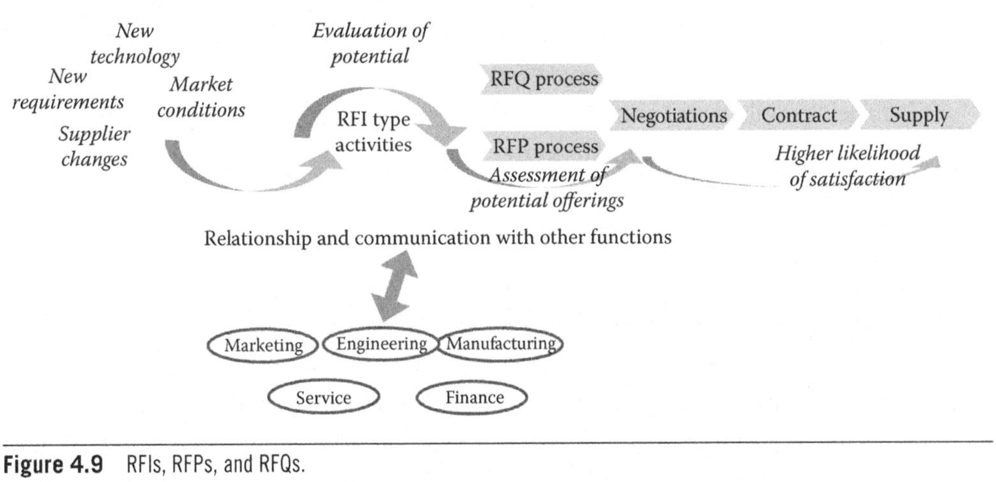

Figure 4.9 RFIs, RFPs, and RFQs.

that we feel there may be potential for a combination of their proposals and then leave them to it. Orchestrating the combination ourselves to any greater extent could run us into trouble if they do not agree and we have spilled the beans on their ideas.

In some countries and depending on the end customer (for example government-owned customers) suppliers can legally ask to see the selection criteria and how they were evaluated. They cannot ask for your evaluation scores for their competitors.

RFPs can enable us to get the best product from the best suppliers at the best cost. To make sure we achieve this, I have found it useful to think about the following questions:

1. *What do I want to achieve?* In an RFQ, I may be able to tie down specifications and other requirements tightly and ask for pricing. This may be OK, but we may miss something new that one of the suppliers can offer. So only do this if you really know the answer.
2. *Who should I ask to bid?* It may be obvious who the potential suppliers are. By all means, get them involved. However, it is always good to see who the not-so-obvious suppliers might be and get some of them involved to pick up what is new, or challenge conventional approaches and especially pricing. Look at smaller players, new players, or some from slightly different marketplaces. Always asking the same suppliers to bid and selecting the same one is likely to cause the losers to treat your requests less than seriously. If you intend to select the same supplier, maybe you can negotiate new prices or other changes directly. A supplier will also know you always choose them, so unless you are open to a selection process, do not ask for bids.
3. *What is the bid process I need to use?* This is not as simple as putting out an RFP, getting the quotations, and selecting the best. Some outline discussions with potential bidders may help to make sure we ask for the right things in the RFP. We may find out about issues which can ensure we get everything we can out of the deal. Initial supplier discussions are discussed in more detail below.
4. *Who should I involve in my team?* You certainly need people who know about the technicalities of what you need to buy. You most likely need commercial people, negotiators, and internal customers. Serious players select and organize their people very carefully in order to support the process. The potential suppliers are certainly likely to do the same. They are likely to have red teams and maybe other color teams to maximize their bid chances. Therefore, we need to organize as well as they do.

Initial meetings with suppliers can be quite advantageous for a number of reasons and potential suppliers will usually be very keen to meet because it is a good opportunity to start winning the business. If they can make things difficult or even rule out their competitors, they will try hard to do so. For example, they may try to get you to include specific requirements in RFQs and RFPs that other suppliers would find hard to meet. They may, sometimes quite unfairly, let you have information against their competitor, although this is a bad tactic in my experience. From our point of view, it is a real opportunity to gain market, technology, and competitive information.

Some RFQs are evaluated and the business awarded, but some require an assessment visit. This may be by one person, but I advise against it. Take at least two; one to talk and one to takes notes and think. Quite often there will be technical specialists on the visiting team to assess an area such as PCB assembly. They will want to look at the equipment, techniques being used, handling, documentation, quality data, and rework facilities, for example. On such a visit, I find it useful to meet some of the management team. These are the people who will shape their ability, performance, and contribution as a supplier. If the sourcing is significantly important, you need to meet and assess the key people.

You will set a timeline for the bidding process. Take into account events and issues such as the end of existing contracts, changing market conditions, and the end of the fiscal year. Time scales are best kept short, but also have to be sufficient for good bids to be developed, questions asked and answered (both ways), and a clear understanding of offers achieved. You will need to allow time for final negotiations and you may want to include a best and final offers (BFO) stage. It is possible to make extensions to the bid timetable. Potential suppliers may also ask for them. You must judge the reasons and benefits for doing so.

There may be some no-bid responses and it is important to find out why these are made. They may uncover an issue you had not considered and something may need to be put right. This may need a discussion with the no-bidder in case the reason put forward is not the real one. Potential reasons for a no-bid could include

- *They do not think this piece of business is a good match for their capabilities.* Did we misunderstand their organization? Is there a problem with our RFQ/RFP?
- *They think they have little chance of winning the bid and do not want to waste their time.* This one may require some action; for example, if they know one bidder always wins.
- *They do not want the business.* Is it too small, not profitable enough, have we stipulated conditions that are too onerous, or are they out of capacity?
- *They are competing with a business owned by the same group.* We need to make sure we get the best bid based on all their capabilities. Perhaps a joint bid is possible?
- *They are subject to political or even criminal pressures.* This is tricky. We may be in a position to take action if we are a large and powerful organization; even so, this may be dangerous. In some cases an RFQ/RFP may even need to be withdrawn.

Your team will need to include a mix of types of people, plus the right technical or specialist skills. For an RFP in a specialist area such as distribution, for example, it will be very useful to have someone on your team with a good working knowledge of the subject. You *must* have someone who knows about your requirements—what products, demand patterns, customers, locations, and other issues. The mix of people types also needs to be balanced. Simplistically, some need to be tough negotiators, some thoughtful, constructive, critical, and so forth.

The suppliers bidding for the work will have a highly professional process and they will also put together a team to help them win. They will probably use a red team to play the part of the customer and critique their bid. In fact, there is a lot of well-publicized work on how best to organize and win bids through step-by-step processes, color team reviews, and so forth. It is instructive—in fact absolutely necessary—for potential buyers to understand the range of bid-winning approaches potential suppliers might be using.

Your bidding process will be running to the timetable but bidders may ask questions for clarification. It is important to do this in writing because their content may materially affect the bid and subsequent business. Unless there are sound commercial or legal reasons for not doing so, I think it is good practice to share answers with all bidders. This is efficient because questions will not be repeated and is promotes a *level playing field*. You should not, however, share any information that is proprietary to any of the suppliers or that can prejudice the bid process unfairly.

Your team will assess bids on the basis of the criteria you agreed on and are likely to have shared with bidders. It is likely that you will have questions to clarify what is said in some of the bids. It is unlikely that these will need to be shared with other bidders if they are specific to a bid. Questions and answers should be in writing and managed formally because they may be material to the awarding of the contract and subsequent business.

After the bid assessment and satisfactory answering of questions, you may announce one potential supplier to be the *preferred bidder*. Final negotiations can be made that may need to clarify some points in detail but will certainly examine further detailed ways to reduce costs and improve the offer. Other bidders will be told about the preferred bidder status and they will usually be held in abeyance until negotiations are complete and the bid moves to the contract stage.

An RFQ is reasonably specific about what goods or services are required. However, a RFP is more useful when you wish to give potential suppliers more freedom to suggest their solution for what you want to do. You need to explain what you need to do and provide any data they need. If the data is complex, a modern way to do this is to make an online *data room* containing the relevant data available to potential suppliers. There are organizations who will help you with this and host the data room for you.

Companies responding to your RFP are providing you with their intellectual property. You cannot pass this on to their competitors or use it yourselves. You would be stealing their intellectual property and they are likely to take legal action against you.

Source on the basis of all appropriate criteria. Use a team where appropriate. Develop and agree a weighted checklist of criteria for evaluation and selection.

* * *

4.6 Pricing

Prices are negotiated rather than calculated by a formula. Even knowing what the right price is can be difficult. The shopping experience in a bazar is an extreme example, but there is always *elasticity* in the first price quoted by even the most professional organization. Figure 4.10 represents a way to start estimating a target price that is at the right level.

The logic of the material in Figure 4.10 is that it starts with the supposition that the supplier's quotation is likely to be the highest we will pay (unless there are some *add-ons* he or she may try to apply). We might equally apply the same logic, but starting instead with the current/previous year's price.

We look for the possible deflationary influences that may enable us to apply a downward influence on the price. Some examples are listed in the downward arrow in Figure 4.10. These fall into different categories:

1. Tangible reductions could be the results of cost reduction work, increases in volume, cost reduction curves—e.g., for electronic components, deflation/changes in currency value. These may even be offered by a supplier, but alternatively, they are reasonably easy to establish.
2. *Influences on price*, such as competitor quotations, market downturns, and price bank comparative data. These may be known to the supplier, who will try to avoid his or her price being driven down by them, but we should be able to research and estimate them to establish a *right price.*
3. *Intangible reductions* due to pressure of negotiations. The reductions will be greater if we research and establish the right price and then negotiate it with skill and determination.

Unfortunately, there is also a list of possible inflationary factors that are shown in the upward arrow in Figure 4.10, most of which are a reflection of those in the

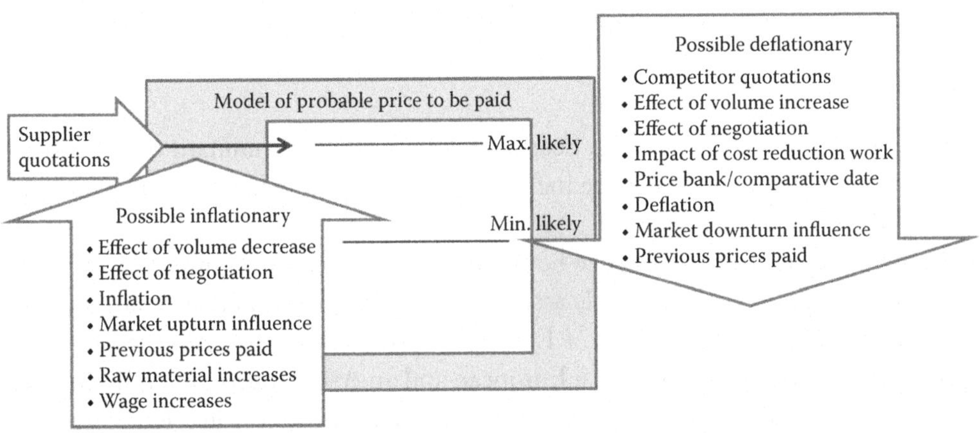

Figure 4.10 Modeling potential pricing.

downward arrow. For example, we may pay a higher price if our volumes are reduced, the market may be moving against us, or raw material prices have increased.

It is our job in the supply chain to collect all the information we can that will influence prices and use it with skill to negotiate the best price.

Figure 4.10 is largely about trying to work out what the *right* price is. If we can negotiate prices down below this level we should. Only where low prices will damage a supplier and consequently our own business should we back away from doing so.

Price banks are mentioned in Figure 4.10 and can provide a very useful guide through pricing being achieved by others. Prices paid for specific items or for types such as steel alloys can be shared. Usually this is across a group of companies or among companies who are cooperating in some way.

4.6.1 Estimating the Right Price for Costing Purposes

Many businesses project the cost of their products using the BOM and standard costs held on the ERP system. In industries with stable costs, standard costs are often set at the average of last year's prices or the last price paid (if prices are falling, the latter is a more prudent approach for inventory valuation purposes and therefore favored by auditors).

In businesses where prices fluctuate more, the supply chain (typically purchasing) needs to estimate the prices that will be paid in the coming year. Even in these industries, there will only be certain components that will vary a lot from the norm, but their price projections usually have to be estimated individually. Price reduction curves are usually available from manufacturers and they can be steep during the periods when the development costs have been recouped and competitor product is becoming available. Precise policy for such parts needs to be agreed on for specific businesses. A typical approach for me has been to estimate the average price in each quarter.

Prices can be surprisingly elastic. Set your targets by properly understanding the issues that might apply pressure both up and down.

* * *

4.7 Total Cost

We used the term standard cost in Section 4.6.1. Usually the standard cost is actually just the price of the item. It may be inclusive of the supplier's delivery costs, but many other costs will not be included.

As an example of the difference between prices and costs, Figure 4.11 compares two possible suppliers. The example assumes we are in Europe and are sourcing some telecommunication equipment for a European installation program. The prices and some additional costs are listed for a European and an Asian source. In addition, there is a planned installation schedule that starts in 4 months' time—just long enough to decide who to place the orders with.

Prices and costs		
	European source	**Asian source**
The price of the part is	$175.00	$120 (30% lower price)
	Shipping included	$25.5 including packing
		Container loads of 200 units
Shipping insurance	Included	$7.80 (special rate)
Duty	N/A	$6.65
First cost comparison	$175.00	$159.95
		(10% lower cost?)
No minimum order size if contract is for over 1000		
Failure rate estimate	200 per million	90 per million
Lead-time	6 weeks	6 weeks + 1 month at sea

Scheduled requirements								
Demand	**Month 4**		**Month 5**		**Month 6**		**Month 7**	**Month 8**
	420	440	410	400				
Possible extension					150	300		

Total costs for 1670 units		
Cost for 1670 units	$292,250	$267,116
ADDITIONAL ASIA COSTS		
Inventory cost while at sea for 5.5 weeks (1670 × 159.95 × 2.44%) = $6517		
Cost to manage shipping (1 person for 3 hours/week at $40/hour) = $600		
Cost of noncontainer load 70 units at $40 = $2800		
Landed cost for 1670 units	$292,250	$277,033
The Asian product now appears to cost only 5% less than the European product!		

Figure 4.11 Example costs for sourcing from Asia or Europe for European consumption.

We can assume the cost of failures is negligible, but we need to allow for inventory holding costs while shipping the Asian product (which has been paid for). An allowance has also to be made for the cost to manage shipping.

When we look at the schedule requirements, we also find they do not synchronize well with the 200-per-container shipping lots. We have therefore added some unit shipping costs for less than a full container load.

In this example, Asia is 30% cheaper on a straight comparison of PRICE. However, the addition of known costs brings the comparison to $292,250 versus $277,033 (for the total project quantity). The Asian cost is now less than 5% lower. Considered on a project cost basis, the Asian supply is $15,217 lower. The advantages of the Asian supplier are rapidly diminishing. In this case, the risks and amount of work needed to get the Asian supply going would mean I would only select them for parts that would be required regularly over a long period.

There is no fixed formula for determining total cost. You will need to think these through depending on the parts and supply situation. This makes the total cost difficult to measure on a completely standard basis. When making the case for a supplier selection on the basis of *total cost* or for a cost reduction initiative, a model of the relative costs is needed. If you need help or you need the support of finance, they may also be able to help you with the model construction and valuation.

Total cost may favor a source whose price is actually higher. Although it may be difficult, evaluate the total cost as well as the price before making significant decisions.

* * *

4.8 Risk Assessment and Mitigation

Supply chains are full of risks. Large or small events can bring enough disruption to the supply chain to stop us shipping to customers—damaging customer confidence and losing us money. Internal and external supply chain issues such as lack of capacity, quality problems, shortage of material, or distribution delays frequently cause problems. Larger-scale problems such as a volcanic ash cloud disrupting aviation or a flu epidemic halting shipments cause more dramatic problems—hopefully less often.

We cannot remove all risks. In fact, if you do, you are not in business. Business is about taking risks. We can, and should, assess risks carefully and take measured steps to mitigate against them or their effects. This is as important in the supply chain as any part of the business, in fact probably more so.

FMEA is a familiar tool in engineering. It stands for failure mode and effect analysis. Figure 4.12 shows an example of a supply chain risk assessment using FMEA.

In column 1 we need to list all the plausible causes of failure we can define. We usually organize a wide range of supply chain associated people to brainstorm these risks. This includes technical, logistics, financial, and business people. We collect their brainstormed risks and organize them, carefully combining any duplicates. We then circulate the risks to the same people, and often following a discussion, we get everyone to give us their assessment of likelihood (/10), potential effect/seriousness (/10), and duration (how long the problem will take to fix) (/10) (this last one is not usually included in engineering assessments, but is a useful addition for the supply chain). We collect the scores, then complete the FMEA risk index column. This is *the Likelihood* × *Potential Effect* × *Duration*, expressed as a percentage in this case. The highest scores are the first priority for mitigation. Usually, these are best addressed by a team of supply chain, manufacturing, and/or engineering people who understand the risks and what can be done about them.

Some examples we know might be useful:

- The sole supplier has a unique dedicated press for one of our essential components. A serious breakdown would take up to 2 weeks to repair. Mitigation—the supplier has agreed to keep 2 weeks of safety stock of the critical component at his premises as part of the VMI arrangements.

FMEA assessment

Product
Product manager

RAG status
Date of evaluation

Potential failure mode	Likelihood of failure	Like-lihood/ 10	Potential effect of failure	Potential effect/ 10	Potential duration of failure	Potential duration/ 10	FMEA index	Mitigation
Adequate carbon fiber not available	Aerospace demand reducing	4	Reduced ability to manufacture	9	May require supplier investment cycle	8	29%	Investigate other intermediate casing materials
No short-term capacity at current machinist	Sector business dropping	2	Change of supplier will cause delay and add risk	5	Three months required to get new supplier running	6	6%	Prebook capacity at current machinist
Chip price reduction curve delayed	General demand dropping	4	Product cost could increase by 6%	6	Potential 6 month lag	5	12%	Ask manufacturer for price support through distributor
Specialist transducer developer/ supplier goes bankrupt	Reports current business down 35% and cash flow difficult	6	Transducer is critical; program could be halted for re-engineering	8	With technology transfer, 6-month delay; 18 months without	8	38%	Investigate purchase of supplier *or* purchase of design IPR

Figure 4.12 Failure mode and effect analysis (FMEA).

- The price of copper may fluctuate so much by the time we need it for the contract we have just won, it may more than wipe out our profit. Mitigation—arrangements have been made to buy forward the required amount of copper, thus stabilizing the price.
- Our supplier for a critical component is small and vulnerable to changes in demand level. We fear he may become insolvent in a market downturn. The option of a second source is difficult because the supplier has special expertise and in any case, volumes are too small. Mitigation—arrangements were made to keep close and watch the supplier's financial situation, if need be, providing him with additional work to support the business.
- An extremely good supplier of intricate machined parts is run by the aging owner with no succession plan and no obvious candidates. Mitigation—an arrangement was made for another good machining business to buy the business, providing a succession plan and reducing the risk to the business.
- The forecast can be wrong by 50% either way on a month-by-month basis and we are in danger of having too much inventory or losing sales (or both!). Mitigation—an arrangement is made with the key suppliers for them to be able

to increase or decrease by this amount within the lead time. This is in exchange for a commitment to buy (although rescheduled) and a long-term contract.

- A flu outbreak in southern China may disrupt shipments into Hong Kong and then to us. Discussions led to the conclusion that the flow of trade through the border was so important that goods would be likely to flow again after a few days, even though people movement may be restricted. Mitigation—no further action was taken at the time.
- The contract manufacturer of a plastic component does not currently have good process control. Quality problems have arisen that could prove serious if they recur. Mitigation—we have temporarily positioned one of our own process engineers in their plant to both check that the process is being run correctly and also teach the supplier the importance and approach we require for process control.

Risk assessment and mitigation is one of the features that differentiate between logistics and purchasing organizations and true supply chain operations.

The supply chain is a risky environment. Form a team to help you consider and evaluate the risks. Risk assessment is an ongoing activity and resources need to be made available to keep it up to date.

* * *

4.9 Relationships

Different types of relationships are required to support what we want from suppliers. Figure 4.13 describes these as

- *Disconnected*, where we will buy commodity items and do not need to work with the supplier at all. Typically they are one-off purchases. These may be candidates for the use of a buy house.

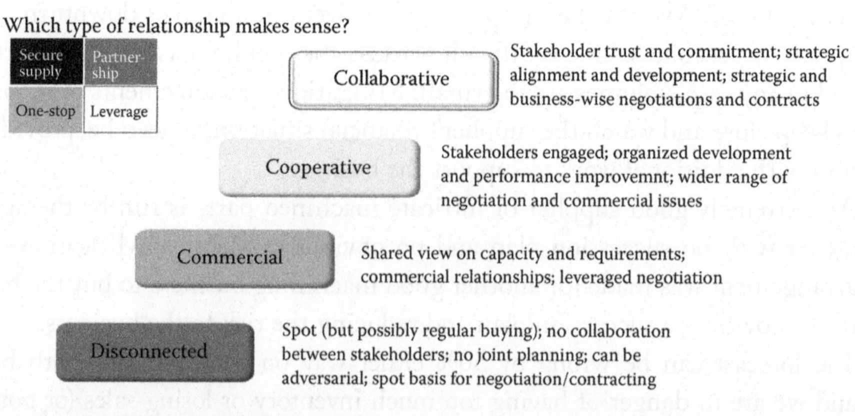

Figure 4.13 Types of supplier relationships.

- *Commercial*, often for commodity buys, where we may communicate and agree capacity and commercial requirements but do not need to share technology or work together to improve service level performance.
- *Cooperative*, where we need to work together on performance improvements and share information. We need to organize people in several areas of the business to enable this to work.
- *Collaborative*, the most sophisticated type of relationship, certainly required in the partnership strategic box. This is often where we are sharing important technology.

None of these relationship types is wrong. They all have their place. In fact, there is no point spending the large amount of time and effort involved in developing collaborative or even cooperative relationships if they are not needed.

Disconnected relationships are not wrong. We do not need a relationship if, for example, we only make a single purchase of a widely available item that is of little consequence (the value is low, there are no quality, safety, or other issues) and then we may never use the supplier again. There is no need to develop a relationship. However, a disconnected relationship is wrong if the purchase is of some consequence—even if it is a one-off purchase. Some form of agreement becomes essential because the parts are valuable or have some intrinsic risk. A written agreement or a contract is needed that makes it clear what the buyer's and seller's responsibilities are. This becomes a commercial relationship.

Commercial relationships are appropriate for many leverage purchases (four-box/Kraljic strategy). There is some form of contract and buyer and seller responsibilities are clear. Ongoing business may be supported by arrangements for scheduling deliveries, managing and returning rejected materials, payment of invoices, and so forth. All the commercial terms for doing business are thoroughly understood and documented. Going further than this by jointly working on some shared objective such as quality improvement will move the relationship into a cooperative one.

Cooperative relationships can also deliver performance and cost improvements through working together and sharing information, targets, and the benefits. For example, a supplier manufactured a desktop monitor for a maker of specialist computer equipment. The monitor was customized, but was based on one of the supplier's standard products. The quality of the customized monitor was disappointing, especially compared with the standard product. The two companies cooperated by forming a joint quality improvement team. Failure data was collected, analyzed, and grouped carefully by the team. It should not have been a surprise that the faults were mainly associated with the special modifications to the design. The joint team used quality improvement techniques and quickly achieved a quality level similar to the standard product. The customer's team members contributed to existing quality improvement expertise. The supplier's team members collected the failure data and did most of the analysis. However, both teams contributed this effort and expertise free of charge. The customer's other options would have been to re-source the monitor or just straightforwardly demand improvements from the supplier. However, they judged that re-sourcing would take time and money and could easily end with the same result because the monitors were customized. It was therefore a commercial decision—the lowest cost and most beneficial route. However, there was a cooperative relationship because

- Both parties put effort and money into the project—a significant effort in the context of day-to-day operations, but not of the level to make a critical difference to the financial performance of the business.
- There was a written agreement, not intended to be a contract to require certain things by law, but to make sure each party understood the work they were committing to.
- There was a mutual benefit from the cooperation.

A collaborative relationship is similar in many ways to what we described above as cooperative. However, there is usually much more at stake. Collaboration is often needed when important parts of a product or service are outsourced. The supply is critical to the customer because it is an intrinsic part of his or her own offering to the customer. To the supplier, the business will be a very significant part of his or her revenue stream and he or she is likely to invest in product design and development, tooling, and capacity. The risks to both supplier and customer are high. Perhaps not large enough to break their businesses, but often such that failure or significant problems in either party delivering on their obligations would badly affect financial results and reputation. In these cases

- We would expect a legal contract to be in place that quantified each party's obligations and deliverables against a timetable, and the levels of finance and returns expected.
- We would also expect a supply chain agreement, perhaps as an appendix to the contract, to describe the practical arrangements for developing the program.
- The work done by both parties will be integrated in some sense. For example:
    - During the formation of the project, data about potential customer orders, project costs, and timescales need to be shared
    - Technical information, drawings, and specifications need to be shared, often including sharing of CAE files
    - Design teams need to share their designs
    - Project management must in some way be across both parties' activities
    - Capacity development must be matched at supplier and customer, and introduction time scales need to be agreed which are achievable by both parties
    - Delivery scheduling arrangements need to be agreed in line with customer forecasts and the capacity of both parties
    - Problem solving arrangements need to be agreed—usually by some sort of joint team—managed by a project manager

One of the characteristics of a collaborative relationship and venture is that there is a high level of mutual risk and gain. Parties are *in it together.* They are mutually dependent. It is vital that senior and key people understand the mutuality of the relationship. They may need to approve money or resources or support a joint project; they may say the wrong thing in a key meeting or otherwise damage the collaboration. It is a supply chain responsibility to make sure stakeholders understand the collaborative relationship and its objectives.

How should a business ensure as much as possible that their prospective partner will be a good one? That they will not merely deliver on their obligations when things are straightforward, but will help to solve the numerous and serious difficulties that often arise. In our experience, many people rely too much on the legal contract to ensure this level of collaborative effort. In reality, situations are often *outside the contract* when they get really difficult. In many senses, if we rely on a contract to ensure that a *collaborative supplier* does what we need, many would say we have misjudged the importance of the relationship. Contracts are not a good way to ensure someone *goes the extra mile.* When things go wrong, we start to rely on the relationship and the character of the organization and individuals to continue, put in more effort, change their plans, and accept lower profits than anticipated as necessary to make the program successful. The supplier also needs to trust the buyer to deliver on his or her commitments.

One of the major Japanese chip manufacturers supplied us with a range of chips, including some to our own design. They also supplied memory chips that were notoriously difficult to procure at the time. They were changing format and capacity on a frequent basis and supplies of old chips were disrupted before the new chips were fully available. We had made a mistake and scheduled an excess of $500,000 worth of chips at a time when their price was falling rapidly. Our business was relatively small and the loss of money would be very significant. We scheduled a meeting with the supplier's head of sales and told him about the problem. Although the chips were flowing into his warehouse and were depreciating in value as we spoke, he agreed to cut the schedule. I do not know what he did with the chips, but my guess is he lost money. Ours was a cooperative relationship rather than collaborative, but he did us a great service. This was certainly not because of a contract; we had been a customer for about 2 years, but not a large one in their terms—I believe they helped us because of the relationship we had built up. It was professional, quite tough, but straightforward.

Trust needs to be earned and developed—and it takes time. If the relationship and the prospects from the joint working look good, then the trust developing work can start. Track record is a more dependable basis for trust than personal judgment or feelings or contracts. Real trust and relationships generally take at least 3 years to develop to the top of the *trust pyramid* shown in Figure 4.14.

Joint initiatives are a valuable way to develop the relationship and trust. In order to be successful, joint initiatives must

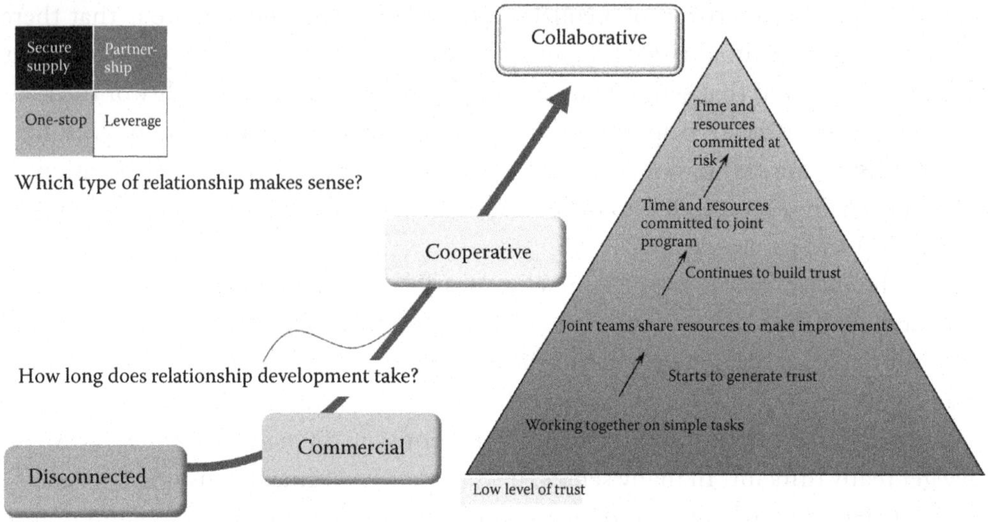

Figure 4.14 Developing trust.

- Be open and honest
- Include senior people to sponsor and guide the work and development of relationships
- Include work *at the shop floor* on practical improvements or developments
- Have clear targets and time scales
- Be clear what is expected from each party
- Celebrate the successes and accept and learn from mistakes honestly

On an assignment with a major contractor, we developed new supply arrangements for a large-scale specialist machined plant. Their approach was to tender all work and select the supplier who was capable of making the parts to a high standard at a low price. One of the major problems (this is quite common) was that engineering would make improvements to the design and this would be used by the suppliers to increase the price. The new approach was to select only four suppliers who would each have a guaranteed minimum amount of work in exchange for a clear basis of price for any parts. The basis of price was related to the material cost, machining costs, overhead, and profit. The contract would last for 3 years.

Trust was developed between the customer and selected suppliers, but also between the four suppliers selected to share the work. The suppliers of these large machined and welded structures were soon assisting with the adjustment of designs to improve functionality and reduce total costs. Previously, they were increasing their profit as the inevitable design changes were introduced. At one stage, one of the buyers spoke to me after returning from the monthly meeting that was held with the four suppliers. He was surprised to see one of the suppliers offering to lend tooling to another for work that was similar to their own! The client's business was required by its parent organization to retender such work every 3 years. The retender was done in an open and straightforward way, and perhaps surprisingly, the cooperative relationship survived the retender; in fact, many of the new working practices were written into the tender.

All internal and external supplier relationships are the responsibility of the supply chain. We are also responsible for supplier relationships with a range of other people within our organization, as shown in Figure 4.14. The majority of supply chain people operate at the middle portion of the diagram—in other words, in some form of management role, either of suppliers or the supply chain. The relationships we need to manage may stretch from supplier's supplier to customer's customer, and from upper management to the tactical transaction level.

The lower part of Figure 4.15 represents the transactional level. This is where schedules are managed, deliveries are made, invoices are sent, and payments are made. Relationships at this level—from the supplier's delivery employees, to goods receiving, to accounts payable—all impact the service we receive. We can often improve things here by doing something simple such as checking why a supplier's invoices have not been paid. These relationships are all ultimately the responsibility of the supply chain. Check and improve them and the service you get will also improve.

A client planned for a member of senior management to visit a contractor in Dubai. The contractor had increased prices continuously and significantly over the previous 18 months. The senior manager intended to, shall we say, *take up the matter vigorously*. Just before he went, a supply chain team investigated the causes of the price rises and found they were due to

1. Increases in raw material price to the contractor
2. An increase in the minimum wage in Dubai that had increased wages generally
3. Engineering changes made by the client

Luckily, the new analysis was early enough to change the approach the senior manager took. He now approached it on the basis of *How can we work together to make*

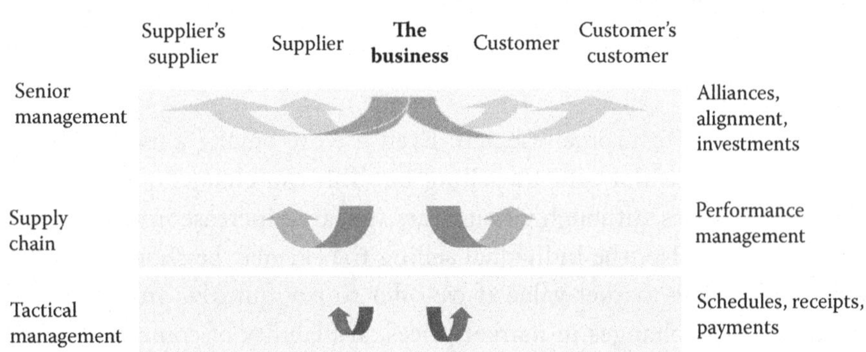

Figure 4.15 Internal and external supply chain relationships.

savings in cost? rather than the one initially envisaged. This is an example of supply chain helping to manage senior level relationships.

During 2009, one client was rightly concerned for the financial health of his suppliers in the economic downturn. As he checked through, the most vulnerable he discovered was a supplier of critical parts to one of his suppliers (hence the importance of *supplier's supplier*). This was one of only three businesses able to make a particular special component. He was in the secure supply quadrant. The client's senior management signed off a support package to keep the small supplier's supplier going through the downturn in order to protect their own critical supplies. They provided the supplier with orders although they did not need the parts. It was a smart move, but senior managers have to be aware of the nature of the supply in order to make such decisions. Making them aware is the responsibility of the supply chain management.

Supply chain managers are responsible for determining and developing the most appropriate type of supply relationship. There are a lot of interfaces with suppliers and they must be managed up and down the organizations and may stretch from supplier's supplier to customer's customer.

* * *

4.10 Negotiation

Negotiation is about agreeing to get our *needs* from another party who also *needs* things from us. There are several good things to bear in mind from the start:

- Negotiating is a process in which we trade needs—theirs and ours. We aim to get more value for us than we trade away.
- If we are likely to work with these people in the future, the way we negotiate should not damage the relationship. In a sense, the whole relationship is negotiation.
- Negotiations are different depending on market conditions, the supply relationship (e.g., Kraljic position), historic relationship, and individual business situations. Negotiations with the same people last year may be completely different this year.

The process starts with some research. Even if we're buying a used car we should research what other similar cars are selling for. This can change quite rapidly if the general market changes, although vintage cars appear to increase in value in any market at the moment! Also, the individual selling the car may be short on cash and be prepared to sell below market value if we offer to pay quickly. This also applies to business deals where changes in market prices, availability of commodities, recently lost orders, or the seller's business strategy could all alter their negotiating position. The pricing analysis described in Section 4.6 may be helpful. This tries to establish benchmarks and what upward and downward pricing pressures may exist. It may help us to understand what level of movement may be possible and why. When negotiating with existing suppliers we should have a lot more insight. We may have information about the health of their business, or we may also be able to leverage any recent lapses in service quality, or offer some assistance from our technical people to help them, for example. We should also be careful to research and understand our own needs. In business, it is too easy to assume that price is the all-consuming need. In reality there may be all sorts of cost, service level, technology development, and quality improvement issues that are either more important than price or should at least go on the list of our needs. Data collection and research is an important part of negotiation—fail to prepare and you prepare to fail! I find it helpful to sketch out a table of needs.

The example table of needs in Figure 4.16 is about a negotiation with a supplier who has just been taken over by a private equity (PE) house. These organizations are about

Their issues and NEEDS	Importance ranking	Our issues and NEEDS	Importance ranking
Their profit margin is 8% (this is below their industry average), and they have just been purchased by a private equity house. Our purchases are 18% of their revenue. *They need to increase their margin.*	1	Our profits are 18% and purchases from this supplier are 2% of our spend. *We need to contain costs.*	3
They have a new technology that they will soon release and want to sell to our competitors and us at a higher price. *They need a high margin on the new technology product.*	2	We would like to grow sales in a market sector that could be driven by their new technology, but our competitors already have a foothold in the sector. *We need to secure supply of the new technology product for the market sector we want to win.*	1
Their production output for their new technology will be limited for the next 12 months due to technical issues. *They need to maximize the return for limited NT product.*	3	*We need enough NT products to satisfy demand in our new market sector.*	1
They have a new competitor who is selling aggressively in the market, but their technology is currently inferior. *They need to rule out their competitor.*	3	We have purchased small amounts from the new competitor and found it to be good but of inconsistent quality. *We need to use their new competitor as a threat.*	2
They produce some *special* product for us that is similar to their newly developed product. It is difficult to produce and the price we pay does not compensate fully for this difficulty; furthermore, margin is only average. *They need to drop this special—perhaps in favor of the new technology product.*	2	*We need either this special product or the new technology version.*	2
They sell a small value of product to our main competitors, but their new technology product could be more interesting to them. *They need high margin sales and selling to our competitor (including the new technology product) could help.*	3	We believe our competitors (who use different technology) would be able to use the new technology product and put us under pressure in our existing strong market sectors as well as those we are trying to enter. *We need to stop them selling the NT product to our competitor—otherwise our entry to the new sector will be difficult.*	2

Figure 4.16 Table of needs.

making money from buying, managing, and selling businesses. The PE house may sell some of the assets of the business, make performance improvements, and usually sell the business for more than it purchased it for. They will encourage their companies to maximize profits and increase prices to customers where possible. One analysis we know they do is to identify customers who are making them too little profit for the effort put in. This leads us to the situation in the table of needs.

The table of needs (Figure 4.16) shows how the situation can start to be quite complex. The supplier in this case will certainly be driven to increase his margin by his new owners. We will want to reduce prices if possible, but the purchases only represent 2% of our spend. The impact is therefore small. Our senior people might tell us that using their new technology (NT) product to gain an advantage over our competitor in our new target sector may be more important. Perhaps we can agree to special access to their new product? They appear to be short of capacity for the NT product in the short term. We need to secure enough for ourselves—especially if this reduces the amount available to our competitor.

We need to think about the information we now have and work on our approach and tactics. The negotiations may not go to plan, but that is no reason not to start with one. We need to make some kind of assessment of their needs and ours in terms of the likely value to us and to them. Price is probably more important to them than to us, although we need to control our costs. Getting ample supplies of the NT product without paying too much is high on our needs list. In the example table of needs, we have started to get an idea of the relative importance of price and availability of an NT product.

There are a number of other interesting ways to analyze our situation that are useful as a structure to think through:

- *Best alternative to negotiated agreement* (BATNA), from the 1981 book *Getting to Yes* by Roger Fisher and William Ury.* This is like a plan B—how good is theirs, how good is ours? In this case, our supplier's BATNA is likely to be to sell to our competitors. Their new technology product could increase their sales. However, it may take some time to replace our volume with that of our competitors. Our BATNA could be the product from our supplier's competitor that we have already tried and found variable in quality. Perhaps if we have the skills and resources, we could assist them to improve the consistency of quality. Both BATNAs have their flaws, but we will each try to talk up our BATNA and talk down theirs: "We have taken some of their product and while I can't say everything is perfect, it is quite an interesting option. We are discussing a joint development project with them."

* Houghton Mifflin, 1981.

- *What is against my interest* (WAMI): For us, our competitors getting the supplier's NT product and using it to make inroads into our market share would be bad.
- *What is in it for me* (WIFM): For us, the new technology product could help us attack our competitor's best market sector. A combination WAMI and WIFM would be to keep all of the new product for ourselves.

Taking into account all our research and thinking through our position and that of our supplier and what options we have, how will we play our cards? Normally, after some formalities, one of the parties will usually set out some sort of opening position. This usually sets out a position in turn that is not the end result they expect. It will be quite biased in their own favor—and quite often extreme. The aim of this is to reset expectations. In many cases, the starting position should almost be shocking. However, it must be plausible; otherwise, we discredit our position from the start. In terms of price, for example, we often want to introduce a price that is lower—and at a completely different level from the one the other party will be considering. The credibility of our opening position could be based on reduced raw material prices, our own competitor's price reductions, improvements in production technology, and so forth. We need to be careful that these are supportable because they will be tested and argued against, and it will count against us if our opening position is discredited.

A few days before the negotiation meeting, the supplier contacts us to say that their CEO would like to attend the meeting and that he would like to tell us about the private equity business that has bought them. They say they are under pressure to increase prices to some customers who currently enjoy extremely low prices.

At the beginning of the meeting, the supplier's CEO says he would like to start by telling us about the purchase of the business and plans. After some information about the new PE owner, the CEO sets out his plan to become the largest supplier of his type of product to the industry, including all our main competitors. He tells us his margins are too low, much lower than ours, and that he is required by his new PE owners to increase his price. He puts forward an increase of 10% saying that is the level needed to meet the PE owner's needs. He is also quite aggressive in putting this forward and says he will drive prices even higher with our competitors. He will also supply the NT product to drive his sales with them, but charge a premium price. His opening position is plausible. It would be possible to argue against the level of increase, but the need for some increase is credible. He has also been quite aggressive in the way he has delivered the message and we decide to open ourselves on another subject.

We announce that we have plans of our own, but his current product does not perform well enough—we need his NT product in conjunction with developments of our own. These would grow our market share and we would be prepared to pay a slight premium for NT, although not as high as he suggests. However, our concerns are about his ability to ship sufficient volume of the new NT product because our demand forecast is high. We give him some tentative numbers and ask him to confirm whether he could meet the demand, including some immediate requirements. The demand we tell him for his NT product is very close to what we believe is his maximum capacity.

This is contrary to the CEOs plans to sell us the old product at a higher price and his NT product to our competitors at an even higher level. His response is that the NT product is at a premium and he can only sell us a limited amount at a 20% higher price; otherwise, he will jeopardize his business plan.

Our BATNA now comes into play and we tell him that our own plans require higher volumes of the NT product. In conjunction with our own technical developments we have also tested his competitor's product and we believe that with some work, this product would work. We cannot tell him his competitor's price, but it is considerably less than his.

The negotiation is not going in the direction the CEO anticipated. He cannot see a clear way forward at this point and returns to the slightly aggressive style again. He says we will be taking a major risk if we go with his competitor's product and that we will fail. He raises his voice and says if we then ask him to supply later, he will charge us at least 30% higher.

Tension between negotiating parties can be very useful. If there is too much tension, participants can sometimes stop operating and thinking properly and no useful progress is made. Too little tension and nothing significant gets done either. Managing the tension to the right level and sometimes even raising it and lowering it can move us forward.

Negotiating teams are sometimes helpful. I like to have a second person on my team if only so that one person can talk and one thinks and writes. Engineers and subject matter experts sometimes need to be on the team because specific issues need to be included. Sometimes a good tactic might be to have our quality manager in the negotiation if we have had some problems. This emphasizes that the problems are important to us and they may be part of the negotiation. All of these issues need to be considered and formed into a game plan for the negotiations. However, it is essential that everyone knows what their role is and sticks to it. In my experience, most people like to get involved in price discussions. Unless this is part of a planned approach, it should be made clear they should leave it to the purchasing people, otherwise, the whole approach might be undermined.

We tell the CEO we have already done preliminary trials with his competitor's product. We ask our technical expert who is on our team to say a few words. These have been rehearsed to some extent and he briefly says they have found with a little work it will work well alongside the new developments of our own. Since the product is also at a lower price, it is an attractive prospect. We tell him there is a possibility that we will enter new markets at a reduced price and introduce new technical features that would increase our market share at the expense of our competitors and his own product. Although we are speaking quietly and politely, this increases his tension further.

The CEO appears to be annoyed, but says nothing. He probably thinks we will not take the risk of changing the product. This is actually true—we would rather stay with his tried and tested product. The CEO also seems to have heard that silence can be used during a standoff in negotiations. We are also silent. This does nothing to reduce the tension. In the end, he breaks first and he says he will not make this offer again, but he will agree to a 15% increase on the NT product, but with only a third of our requirement. The remaining standard product will be supplied at a price increase of 6%.

We say "No." This would clearly not work for us. A further silence follows. At this point it is important not to lose our courage, as we say in the U.K.—to be determined, resolute, even ruthless.

We are the first to speak this time. We say we would, of course, ideally like to continue to use their product. However, the level of demand included in our business plans means we need him to commit to our full forecast demand for the NT product (which we know will be his full capacity for the short term; we are effectively taking all the production of the NT product, which will mean our own competitors will get none). We will pay a 10% increase on price for the NT product and a 4% increase on a smaller quantity of standard product. We attempt to close the deal by suggesting that this meets his needs and gives him a sufficient increase in his profits if he passes similar price increases on to his other customers.

The CEO asks for an a brief adjournment at this point and when we return he has clearly accepted that we should get all of the production of his NT product, but we end up making a small additional increase to the NT product pricing.

The new pricing enables the CEO a good chance of making the improved profit required by his new owners. We have secured all the production of the NT product for the coming year, effectively excluding our competitors. We will pay 12% more for the NT product that we believe we can sell at a slightly higher price with our own developments, even into the new market segment. This is a win–win result.

Figure 4.17 summarizes the negotiating process.

"Maneuver" is included in the five Ns in Figure 4.17. This describes an unexpected change or new scenario, usually made as an attempt to disrupt the other person's thinking. One of the best we heard of was during the negotiation of the purchase

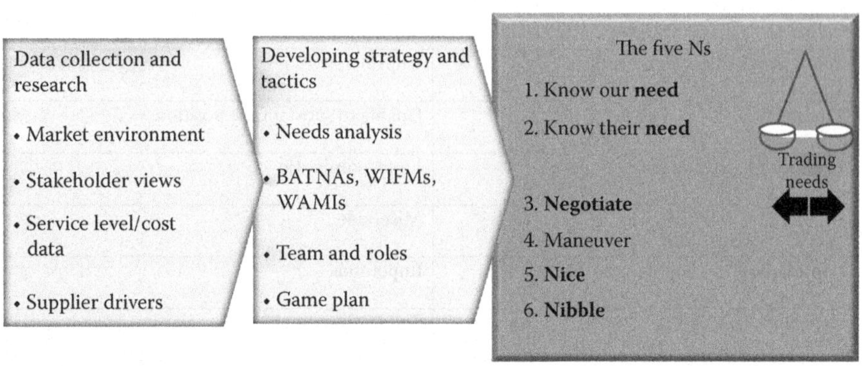

Figure 4.17 The five Ns for negotiation.

of some aircraft. The price of jet engines was about to be discussed and the buying team had been preparing for this very important part of the proceedings. They were surprised when the selling team announced they proposed to "give" them the engines "free of charge." When faced with such a maneuver, do not instantly grab it. There is obviously a catch. In this case, of course, the catch was the maintenance contract, on which the vendor could make more money than from the engine purchase.

We include the word "nice" in the five Ns to emphasize that being disrespectful or rude is usually counterproductive. Even in a situation where someone is being disrespectful to me, I find the best response is to refute their point very politely. In fact, a broad smile and a polite response along the lines of "I can understand why you might think that, but …" can often be disarming to an aggressive approach. This is a matter of personality and people develop their own preferred style. I have found that politely, but firmly setting out my case, backed up by plausible reasoning, works best for me. Persistence too is absolutely necessary. It works best for me if I can support my case with a strong logical justification.

Nibbling refers to further concessions you can win after the main deal has been struck. They may be smaller, but can still be valuable, such as keeping some safety inventory. My wife is good at this. If we are buying me a new suit, she always follows up with a request for a free shirt—it matches the color of the suit—or a free tie. Nibbles can sometimes be quite substantial if they are appended to the main deal: "I think we could agree to that if you could do something to have some inventory available to us on a call-off basis…."

Negotiation requires a delicate balance of personal skills. Some people have more natural ability for negotiation, while others have to practice and work hard at it. Figure 4.18 describes the delicate balance of attitude and behaviors I think work compared with those that lead us into poor negotiating and results.

✓	✗
Open-minded, quick thinking	Fixed ideas, intransigent, closed mind
Polite and firm	Aggressive and rude
Ruthless, relentless, determined	Unprincipled
Plausible	Unable to substantiate position
Calm, quiet, determined	Loud, bombastic
Smiling	Miserable
Considered	Impetuous
Listening	Talking too much

Figure 4.18 Negotiating attitudes and behaviors.

4.10.1 A Negotiation with a Change of Scope

Our supplier of garden and building maintenance services is not performing well against contracted service levels. He has also said he needs to increase his prices because he has had to pay his employees more. We have been in discussions with alternative suppliers who will undertake the work set out in the service agreement at a lower price. His contract is due for renewal and we start a negotiation meeting with him expecting to conclude that we will not renew his contract.

We start the meeting by saying that we are disappointed that he has not performed to contract and that we cannot accept the price increases he has proposed to us ahead of the meeting. It is likely that we will not renew the contract.

He asks for a few minutes to explain some of the details of why his service levels and prices are what they are. He presents information about the work undertaken in a straightforward and factual way. This indicates that among the work they undertake on standard and urgent work requests, urgent requests have increased to four times the level they were when the contract started 3 years ago. He shows us that these are mainly due to three main problems:

1. The heating and ventilating system needs to be cleaned and adjusted on a frequent basis. He says this is because the fuel now being used is not of the same type and contamination causes problems with the burners. A relatively small investment in a new burner should solve these problems.
2. The parking garage is regularly overflowing with cars parking on a grass area, which has resulted in the collapse of some drains and part of the grounds flooding after rain. He suggests, if we must park on the grass, the installation of new concrete pipes along the drainage path close to the edge of the garage.
3. A piece of equipment in one of the offices that is not his responsibility (it is on another maintenance contract with the supplier) has a fault that causes an electrical circuit to trip, cutting off the electrical supply to the office.

Work related to these problems accounts for nearly 70% of the urgent work requests. This urgent work has been done, but at the cost of poor service levels on other regular work. Additional labor could be employed to clear the regular work. However, the level of payment allowed by the contract is fixed. Additionally, he has increased the hourly pay of his employees because of local people shortages.

This leads to the second issue, which is that he has agreed to an increase of 3% with his employees in order to retain staff. He is straightforwardly asking us to cover this with a 3% increase in contract fees. He is also asking for urgent work requests to be billed separately.

Our views have been changed to some extent by the straightforward way he has talked us through these issues. However, we cannot afford a 3% increase in his maintenance contract. We actually need to reduce these costs. Is his reporting of the causes for service level failures correct? Would the alternative providers fair better with the problems described?

After some checking of the facts the supplier presented, which proved quite factual, an agreement was made to make investments in a new burner and the recommended concrete drainage. An extension to the existing contract was agreed without any increase in payment levels. Over the extension period, the maintenance provider agreed to look for ways to increase efficiency in order to offset the wage increase.

In many negotiations, particularly for services, the devil is in the detail. Prices and costs are to a great extent the result of the work involved. Changing the work, solving problems and removing work that should not be needed, and increasing efficiency are at least as fruitful as hard negotiations on the prices and profitability of suppliers. The difficulties unearthed during the negotiations should have been known earlier through good supplier management. However, they were not. There are many cases where negotiations do not go to plan. They uncover important, relevant information previously not known. It means we need to listen sometimes rather than attribute stories of difficulties to whining.

4.10.2 A Negotiation Where Price Is the Dominant Issue

We manufacture a specialist mobile piece of equipment that uses a large piece of electrical equipment made by a supplier. The electrical equipment is mostly standard in the marketplace, but with some minor customization to make it suitable for our product. It represents about 10% of the cost of our product.

Our sales are growing gradually and we are the largest supplier in the European and U.S. markets. We are acceptably profitable, but a serious competitor is offering a similar product at a lower price. We would like to hold our prices firm but cut our costs, and our target for this assembly is 2%. There are no technical or quality problems in this area.

Our competitor has a slightly different design and uses an electrical assembly from our supplier's competitor. We believe their prices are similar.

Our supplier has let it be known that he is looking for a price increase and 7% has been mentioned although we believe his target is more like 3%. We understand their business is quite healthy. Their reported profits are in the upper quartile for their sector and business type. Our product represents nearly 10% of their sales—so quite significant.

We try to define the NEEDs as

OUR NEEDS	SUPPLIER'S NEEDS
2% price reduction	3% price increase
Continued supply (but plausibly from another supplier)	Continued business from us

This is really about price. Unusually, there are no other issues.

We start the negotiation with two purchasing people in a meeting with their two sales people. They have all met before. Each of us gives a short introduction including the state of their business, their outlook, our business together, and current objectives. This leads to an opening position where we need them to reduce their price by 2% and they need to increase it by 3%. Except of course we have asked for 5% and they 7%.

Both of us back up our need for price movement with plausible logic about the target profitability for our businesses as far as shareholders are concerned, competitor positioning, and market situations. During discussions about competitors, we say that we could switch to their competitor's product, which is equivalent and that we believe we could buy more cheaply. Our engineering costs to do this would be quite modest. The discussions remain polite, but we raise the tension by mentioning work we have done to look at the competitor's product. The technical changes are quite minor. We could switch suppliers quite quickly.

Tension is useful in negotiations. As mentioned previously, too little and little will be achieved, but too much is likely to cause people's ability to think to become impaired and negotiations may break down. If you watch some of the political interviewers you will see tension used as a technique. The interviewer will move the tension up and down to put pressure on the politician (they deserve it and it makes good television!). Figure 4.19 shows an optimum level of tension to achieve the most effective results.

The supplier predictably says we would find that his competitor's product will not work as well for us. He gives us a summary of the (admittedly good) service level they have provided and some instances where they have helped us out by shipping within lead time.

We acknowledge this but also point to the price rises we have paid to them in the last 2 years against a backdrop of little inflation. We have maintained our customer prices over this period and have maintained our margins through improvements in efficiency.

There follows some more counterpositioning, where we are both polite but firm. Although we raise and lower tension and can see we are wearing him down, we are not progressing much. The supplier then suggests (this is a surprise) that as we both want prices to move in the opposite position by roughly the same amount, we should agree to keep the existing prices. This is a sign of weakness on his part.

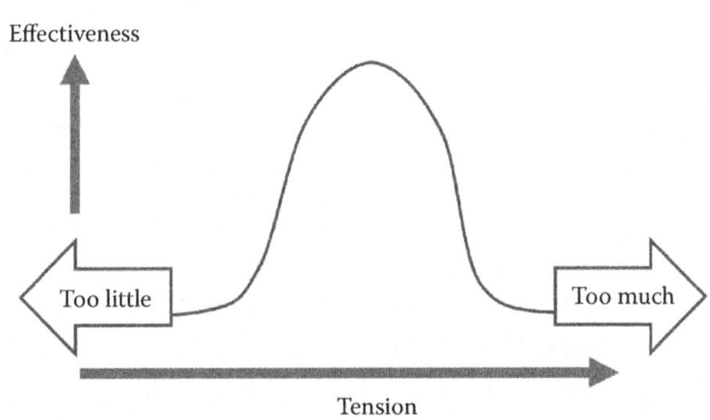

Figure 4.19 Tension in negotiations.

We now say that we would like our business to continue to be at least as profitable for him, so maintaining prices would be a start. However, we use the example of our own efficiency improvements and suggest he could commit to making our 2% price reduction in two 1% stages, based on efficiency improvements himself. Are there ways he could be more efficient? Of course there are. We suggest that we could give him some help starting with joint visits by our manufacturing teams.

After more discussions about the details, during which he commits to timing on the two 1% steps, we agree to a plan. We have a whiteboard in the room and we capture the key points on the whiteboard.

We are not expecting this news to be well received by our supplier's team. Perhaps, at first, this is accurate. However, we follow up quite quickly by inviting his manufacturing team over. Our team gives them a tour and a presentation about some of the improvements completed and underway. We are careful to do this in a constructive way rather than as an ego-boosting exercise. The two teams get on extremely well and start to cooperate in a positive, value-adding way.

4.10.3 A Negotiation with a Surprise Result

We are a large designer and manufacturer of equipment used in mining and oilfield exploration. We are highly profitable and our share price depends on continuing success. There has been a sharp downturn in the market that has reduced our sales drastically. However, we have some long-term customer contracts and have made special deals, so our revenue has only reduced by 25%. We have cut our own resources in terms of direct and indirect labor and some capital investment. Our cash position is strong, but we do not know exactly how long the downturn will last.

Our supplier had (with our assistance) designed and has manufactured a highly suitable subcomponent for our own product for the last 5 years. The design is better than the competitor's and quality and lead time are also good. His price (and indications are his profit margin also) are quite high. His reported profit was 12% (ours was 17%). Our competitors also buy from this supplier.

Based on our research and analysis, the following shows what our needs are

SUPPLIER'S NEEDS	OUR NEEDS
Short term: Generate cash, continued revenue level	Short term: Negotiate cancellation of orders
Retain other customers	Medium term: Reduce prices
Survive	Safeguard access to best-in-class technical product

Our BATNA is not strong. We prefer to buy from this supplier because his design is better than his competitor's. Without his supply, we may need to develop and manufacture our own design. His service level and relationship with us has been good. Based on our experience with them, they are a high-quality business with good management.

The supplier's BATNA is quite strong because he has other customers, although he needs to keep our business because his profit and particularly his cash are down.

Our Plan for Negotiation

- Open with a discussion about how we are hurting in the current environment. We are having to cut costs to retain revenue and we need him to reduce prices. We also do not need all the product he is expecting to make for us in the short term.
- We expect similar comments back from the supplier about his reduced level of business from ourselves and our competitors. In fact, he tells us about the layoffs and cancelation of nonessential expenditure they have made.
- After some discussion we hope to be able to propose and negotiate some reduction in schedule, some reduction in price, but faster payment of invoices to help with his cash flow.

A Maneuver?

- However, at this point, he introduces what can almost be described as a maneuver, except that it appears to be completely factual and quite important to our negotiation. He tells us that if he accepts our proposed reduction in price and schedule, and does the same for our competitors, they may have such bad cash-flow problems they may become insolvent. We had not expected his situation to be quite as bad as his report and so we continue to ask some questions about how bad and how imminent this problem is. He reports that even with the cuts they have already made, his cash flow could be projected to be dangerously low within the next quarter.

A Recess

- Given the unexpected news, we ask for a short recess to discuss our position. During this we conclude that our original plan to cancel some of the schedule and negotiate a lower price because he needed the business was no longer viable. Continuing with the existing schedule and prices would not hurt us, but the possibility of the supplier becoming insolvent is of greater concern. One of our team suggests that a possible solution would be for us to buy the supplier's business. It might make sense because we have a strong cash position, we would gain an advantage over some of our competitors, and we would safeguard an important supplier.

Upon Resuming

- We ask for further details of their financial situation that we can discuss with our senior management, but stress that the situation is also very serious for us. Having got some good data (somewhat unofficially), we call an end to the meeting and ask to schedule a further meeting in a few days after we have met with our own management team.

Upon Returning and Meeting with Our Senior Management

- We stress to our senior management team the importance of the competitor's product to our own, the supplier's good level of service and quality, and provide the best level of financial information we have regarding his cash-flow situation. We also tell our management team about our competitor's reliance on this supplier. We have developed several scenarios and likely results. We present these, including the suggestion that we purchase the supplier's business.

The purchase of the supplier's business is accepted as the lowest risk, potentially achievable, and with the potential to give us a significant advantage over our competitors. We request a new meeting with our supplier, including some of our senior management, and the proposal is made. In fact, this resulted in the purchase of the business and put our business in a much more competitive position.

Some books on negotiation that I have found useful are

- *Getting to Yes*, by Roger Fisher and William Ury (ISBN 01401.57352)
- *Everything Is Negotiable*, by Gavin Kennedy (ISBN 978-1847940018, Random House, 2008)
- *The Art of Negotiating*, by Gerard I. Nierenberg (ISBN 978-1566198165, Barnes Noble, 1995)

Good negotiation is about thorough preparation and understanding the supporting detail as well as being persuasive in the meeting.

* * *

4.11 Supplier Development

Effort spent developing suppliers is usually worthwhile in all the four-box (Kraljic) quadrants shown in Figure 4.20, with the possible exception of leverage. Supplier development can mean a wide range of work that results in product or service levels better meeting our needs. This could be in quality improvement, cost reduction, capacity improvement, lead-time reduction, or product range and type and variant development, for example. In the partnership quadrant particularly, we sometimes joint fund programs with the supplier, we could share resources with them in some way, or do a joint project. Relationships are likely to improve as a result of the work if it is managed well. Without supplier development, we are at the mercy of commercially available products and services. This is therefore an important way in which the supply chain can add value. However, all of these initiatives involve the investment of our time and money and those of the supplier. We need to select what we do carefully.

Partnership suppliers are most likely to benefit most from joint improvement efforts because the monetary value of the purchase is high and in some way difficult, probably because of the technology involved. However, most partners are also very

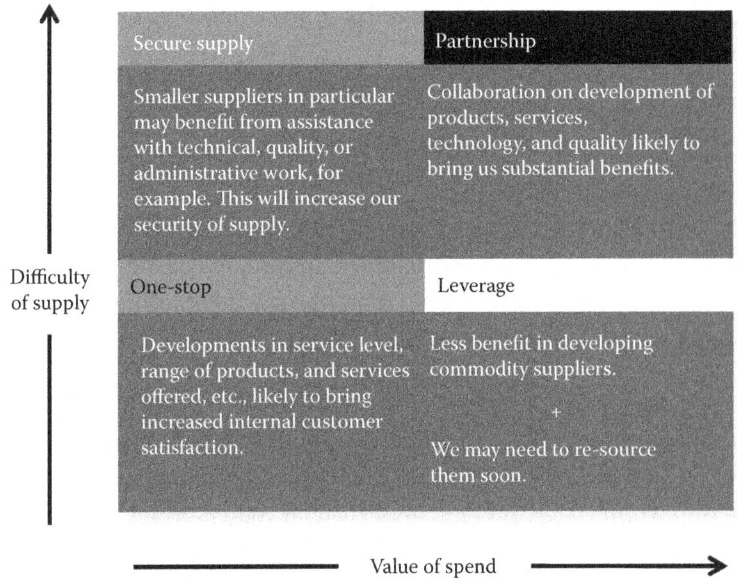

Figure 4.20 Four strategic boxes for supplier development.

competent at their business. Working together is often a joint development of new technology, a new process, or entry to a new market. This is often more like a joint venture than a development of a supplier. However, I will use the following example to highlight that there are risks involved. One of our clients decided to outsource the assembly of a large product to a supplier. The supplier selected is very competent and a well-known company. Joint investment was agreed for a new assembly and test facility, and, managed by the supplier, was opened successfully on time. The facility makes the product to the target cost and quality levels. However, there is a problem with the ability of our client to forecast demand for the product accurately. Large fluctuations in demand occur and the product is complex and costly, with a long lead time. This causes significant difficulties between the two parties, often tying up substantial amounts of cash. This strains relationships in what is otherwise a well-run operation. This example of a well-executed technical project that is hampered by subsequent commercial problems is not uncommon. It should not dissuade us from doing such projects. However, risks such as forecast errors, technology change, competitor developments, and the like should be discussed and factored into arrangements from the start. Developments in the partnership area are likely to be at the highest risk.

A client who makes specialist printing equipment became aware of a revolutionary process that had the possibility to elevate their own product to one of technical leadership in the market. The process was the brainchild of a recent graduate—a friend of one of the new engineers now in our client's organization. Although the process was sound and he had the necessary machinery to make it work on a small scale, our client's supply chain recognized the weakness in his small organization. A number of options are available:

1. Write off the opportunity because the supplier is not sufficiently robust
2. Buy the rights to manufacture from the business and make it an in-house capability
3. Support the graduate's new business with our own cash and people resource to enable him to create a business capable of supplying our future needs
4. Arrange for another company who is a current, capable supplier to purchase the graduate's business and develop it with him

This is clearly a significant business decision involving important issues in engineering, marketing, finance, and manufacturing. The supply chain alone should not make the decision. Option 1 seems to represent a missed opportunity, but if we have other equally important developments underway, it may be the best. Options 2 and 3 also depend on what else we have underway in-house. Do we really have the resources to support the development of a separate business? Option 2 also raises a key *make vs buy* question. The fact that the graduate (not ourselves) has made the development so far might indicate this is not our key area of expertise. Option 4 could be a lower-risk option, but it would require some delicate organization and a sound contractual agreement about future product availability to our business. The supply chain should certainly manage a process, including senior management participation, to evaluate the options and take the decision. The supply chain is also likely to be responsible for managing the implementation of whichever plan is chosen. Options 3 and 4 are quite formidable supply chain and business tasks, but they are the type of actions that can enable businesses to become leaders in innovation.

Secure supply types are most likely to benefit from our help because they tend to be smaller businesses. If we have good capabilities in technical, quality improvement, or planning, it may be to our own benefit to give them some free help. The improvements themselves are likely to improve our security of supply. One client we know well has a good track record of quality improvement using a well-established continuous improvement approach. A supplier of a significant subassembly of their product was delivering a product with a high rate of failure. Our client supported the supplier in running a quality improvement program using a joint team. The client's quality expert led the supplier team through a process to analyze product faults from a 12-week period and diagnose the root cause of the failure. The highest occurring and most easily addressed root causes were then solved one by one. Solutions included operator training, adding features to assembly jigs, some small design changes, and additional inspection steps. Improvements in quality were almost instant as each improvement was implemented and tested. After a period of only 6 months the supplier was running his own quality improvement teams and faults had been reduced to 0.1% of receipts.

In the one-stop quadrant, pricing levels are low and the product is standard. However, there are potentially very valuable reductions in administrative costs to be made. Improved customer service levels are also likely to be financially worthwhile. Users of the one-stop services are the most likely people to identify beneficial improvements that could be made with the supplier. One of our clients operates an engineering maintenance and support facility in China. The facility uses a wide range of spares that are supplied from the manufacturer. However, a large range of tools, materials, and standard parts are also used. These can be sourced locally. However, clearly tired of the nonvalue-adding administration concerned with the purchase of these parts, the head of operations set up an arrangement with a local supplier to open a "store" within the facility. The supplier set up the store to supply the most commonly used hand tools and materials. These were issued at a hatch by the supplier's staff to the client's employees who presented a valid, approved requisition. The supplier kept a spreadsheet record of all issues of parts and materials. At the end of the month a copy of the spreadsheet and a single invoice were sent to finance for checking and payment. This was so successful that the range of items was gradually increased to include more hand tools, safety equipment, thread locking compounds, coveralls, and even stationery. The resulting level of administrative cost is now very low and the service level provided is excellent.

Developments on the leverage quadrant need to take account of the likelihood of re-sourcing; otherwise, development work might be wasted. However, we should still agree and even *require* developments and improvements. They are likely to be of a shorter time scale and require the supplier to achieve the development themselves. Three-year contract arrangements are common where we are trying to get leverage suppliers to undertake improvements for us. These might enable investment in better tooling, for example, which might improve quality and reduce costs.

Depending on the level of spend and type of purchase, improvements should be driven by commodity managers and supplier managers. It is a fundamental part of these jobs to be communicating with stakeholders to make an assessment of what improvements in supplier performance are needed most.

Deciding where to start with supplier development is dependent on business priorities. The obvious truths are

1. If cost down is a really big priority because you are uncompetitive or making poor margins, you clearly need to start with leverage and partnership suppliers.

The quickest results may come from straightforward leverage rather than true supplier development. An analysis of spend across the supply base is a good start to see what scope there might be for rationalizing or re-sourcing—a possible way to quick wins. This could be as crude as consolidating requirements to fewer suppliers, asking for lower prices or, declaring "regrettably, we will need to drop them as a supplier." That is what the leverage quadrant is about. Even if you have been making good purchases and getting low prices, this is always worth looking at. Something might have changed.

Next, it is always worth discussing cost-down opportunities with suppliers, even in the leverage quadrant. Building maintenance contractors, machine shops, printers, caterers, and warehousing and distribution providers are all leverage suppliers who often have very good ideas about how prices or costs can be cut. Machine shops will often be eager to tell you about features of your design that can be achieved with easier and less costly machining operations. A review of the detailed work being carried out by service-level providers will usually uncover things they do that are not required.

For suppliers in the partnership quadrant, the approach needs to be different. We should make it clear to these suppliers that cost reduction is a serious issue with which we expect their help. However, we will be asking for help rather than demanding fast price reductions that will reduce their profits, as in the leverage quadrant. The starting point could be similar to the leverage examples—changes to detailed design and scope of work. They may be able to see this as a sort of opportunity that would bring cost savings. Beyond this, we might decide to use Lean tools to start looking at waste or where cost is added without value. This sort of approach is *expected* by the Japanese auto makers, one of which expects a regular level of employee suggestions to be implemented within 3 months. We know of a major supplier of aerospace equipment that commits to 4% year on year cost reductions to be passed on to their aircraft maker customer. This is an indication of the long-term commitment needed to reap the full benefits of partnership suppliers. Generally, we should be looking for price reductions from our partners through reduction of their costs, not by reducing their own profits (unless they are excessive!). There are likely to be good opportunities to set up joint teams to address these opportunities that will help with relationship and trust development.

2. If the business priorities are about lead time or quality, the analysis is likely to highlight partnership and secure purchases as the likely quadrants where we can find improvements. Discussions similar to the one described above for partner cost reductions are fitting. Section 4.14 discusses how lead time can be cut, which is mainly done by reducing queuing time. This is likely to require detailed discussions and the implementation of new scheduling arrangements with suppliers. This will almost inevitably include risk-taking on inventory. Partnership suppliers are likely to be able to cope with such risks themselves. However, smaller secure suppliers may need help. This could be done, for example, by implementing customer-owned VMI.
3. If your supply chain people are too busy on administration for low value orders, you should look at the one-stop shop quadrant suppliers.

Implementing some form of internal one-stop shop, vending machines, or some other form of VMI should drastically reduce administration—and at potentially higher service levels. Vendors can often implement the solution you need without increasing their price. They are also a useful source of suggestions for managing these types of purchase.

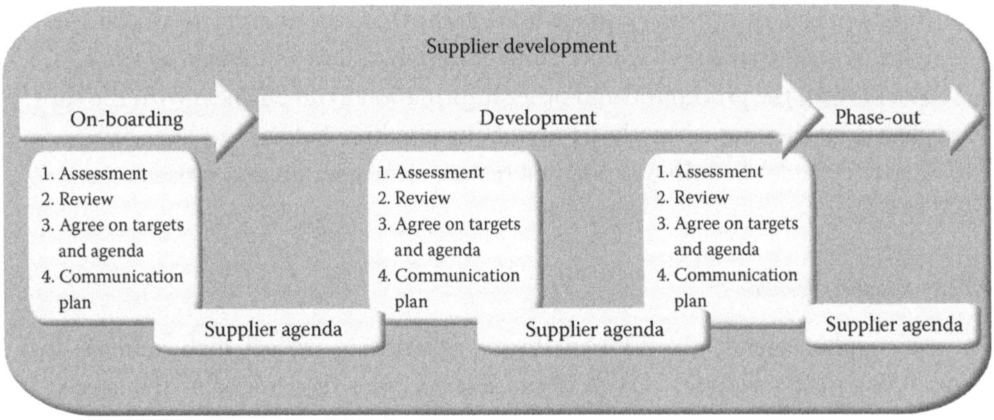

Figure 4.21 Management and development through the life of a supplier.

4. If there is no single imperative area of business improvement requiring focus, discussions with stakeholders could be a good way to make sure we, the suppliers, and our developments, target the most beneficial areas for our business.

Supplier development should start as soon as we decide to source with them.

4.12 Supplier Life Cycle Management

Opportunities to improve supplier performance begin at the sourcing stage and end with good management of the exit of a supplier. We have seen a range of client arrangements for managing suppliers through an *on-boarding* process. We have seen far fewer who have a formal *phase-out* process. The example summarized in Figure 4.21 combines a number of the tools and techniques.

On-boarding a supplier is all too often a matter of signing the contract, specifying what we want, inspecting initial work, and setting up purchase orders and schedules. However, using the more comprehensive checklist below is likely to prevent some omissions and identify some areas for improvement.

On-boarding checklist:

1. Relevant improvement areas from selection assessments are carried through into the agenda
2. First-off components and service inspection issues actioned and now satisfactory
3. First-off shipments pilot (using the same packaging, shipment routes, customs clearance agents, etc.) has been carried out and issues actioned *or* first-off service pilot carried out using live callout, transportation, and equipment and personnel
4. All equipment or tooling issues concluded
5. All pricing issues concluded
6. All contract issues resolved and signed
7. Communication personnel and contacts and procedures concluded (callouts, work schedules, quality, design, contractual, performance, and other issues set up)

8. Supplier has all necessary procedural and other information to do business with us (e.g., they know who to contact for what, how to invoice us, etc.)
9. We have all the procedural and other information to do business with the supplier (e.g., they are correctly set up on our purchase ledger)
10. Capacity and scheduling arrangements are clearly set up and satisfactory

4.12.1 Supplier Agendas

We use supplier agendas in different forms as working project management documents. We call it a supplier agenda because it becomes recognized as the agreed list of requirements and actions that need to be achieved. Sometimes the supplier agenda is just for one supplier. For more ambitious developments it can manage and progress actions by a number of suppliers (and also ourselves). A good example of this is a lead-time reduction program we were involved with that required actions from a final assembler, three suppliers of components and raw materials, and our client (their customer).

The supplier agenda needs to include

1. The objectives, targets, and time scales of the improvement plan.
2. Key performance and improvement measures and targets.
3. If necessary, the equitable share of benefits to be achieved.
4. The actions required by all parties (similar to or may actually be a project plan).
5. Accountabilities and responsibilities for the improvement tasks.
6. Team details (likely to be from supplier and customer, plus others). In addition to the customer team members merely being contacts, many customers supply help and expertise to their suppliers. For example, a major Japanese car maker has a customer support team in North America that includes a large number of engineers dedicated to assisting suppliers. The benefit of this type of investment is to accelerate and increase the improvement of supplier performance.
7. Budgets agreed on and allocated.
8. Agreed-on resource levels from suppliers and customer.
9. Communication plan for initiating the plan, progress reporting, and problem solving.

Good management of a supplier agenda will deliver performance improvements for the customer and the supplier. Trust and relationship benefits will also result.

At or before the conclusion of a supplier agenda, further assessments should be carried out and a fresh supplier agenda agreed on.

It is likely (although not imperative) that the supplier manager will be responsible for the management of the supplier agenda. It is likely that a team combining commercial, technical, manufacturing, quality, and engineering people from both supplier and customer will be involved and they and their actions will need to be managed.

This requires good people and motivational and team-leading capabilities. If necessary, send your supplier managers to an appropriate training course and potentially get them some coaching support for the role.

Supplier phaseout is often a time lacking in positive or motivational activities. However, phaseouts need to be completed professionally because lack of attention to details can cause short- and long-term problems. A supplier agenda and a phaseout checklist are important tools at this stage. Here are some examples of the types of issues that the supplier manager must agree to and correctly manage during phaseout:

1. Reasons for phaseout are properly understood by all stakeholders
2. Person accountable for managing phaseout
3. Phaseout plan with time scales, product volumes, and required actions, accountabilities, and responsibilities
4. Contractual issues are in place before phaseout (e.g., continuing nondisclosure agreements)
5. Schedules for supply to phaseout are agreed to
6. Provision for failures during and after phaseout are agreed to
7. Provision for supply of spares and support are agreed to if necessary
8. Arrangements for ownership or for destroying or maintaining tooling are agreed to (note that if the supplier is to keep the tooling, he or she is likely to charge for storage and maintenance. If the supplier changes machine tools, the tooling may no longer be usable)

The development of supplier capabilities can be one of the most beneficial ways to spend your scarce company resources.

* * *

4.13 Cost Reduction

Prices are usually the largest part of the cost associated with supplies. This should be the first area to think about. Price reduction is not always a matter of pressuring the supplier into giving away his or her margin and negotiating a better price. Some of the supplier's price may well be for features or services that we do not actually need. Even when prices are market-led it is necessary to look at supplier prices and exactly what we are getting.

When supplier prices are said to be *cost*-driven it is legitimate for us to ask the supplier to help us understand the most significant areas of cost. The term *open book* is used a lot when it comes to understanding supplier prices. In my experience the open book usually has one or two *missing pages* or even entire sections (such as overheads that are oversummarized). I do not blame suppliers for this at all.

Until trust can be built, it may be better to start with some joint work—process mapping, for example. This is likely to uncover some cost reduction potential; for example, areas of overlapped cost where we and the supplier both carry out the same tasks. In addition to uncovering unnecessary costs, when managed well this sort of activity can really improve communications with suppliers. One company discovered that parts being made for them in China were being inspected three times. With only a small amount of investment, one inspection achieved much better results at a lower cost. However, the improved communications and understanding uncovered several slightly misunderstood technical specifications. When these were properly understood, the manufacturing methods were simplified. This released even greater cost savings.

Examination of areas outside the supplier's own organization such as raw material and transportation can be a good place to start because it can be nonconfrontational. In some cases, often for raw materials and components, where they have higher volume requirements, a company can pass discounts down to suppliers. Sometimes this can be arranged with just a few phone calls and letters. Sometimes changes in supplier or specification may be needed. Many organizations group their transportation of incoming goods together to reduce costs and also gain better data and control. One company in North America has arranged collection of all goods from U.S. suppliers by one of the large logistics companies. This has saved in the order of $15 million each year. In addition, the data and analysis provided by this highly professional logistics specialist shows pickup frequency, location, lateness, transportation time, and any problem reporting such as poor packaging. Many other businesses do this by getting one supplier to collect and deliver from the other suppliers in their region.

Getting into the areas of the supplier's own value-adding activities can begin with looking at areas of our requirements and specification that drive the supplier's costs up. With some trust being built up, suppliers are likely to start to share more of their direct cost elements. This can lead to understanding of labor rates, allowances, setup times and costs, and so forth. The supplier's overheads should now be clearer and we may then start to ask questions about them and eventually his or her real level of profit from our business. Open-book information is what we need, but we (probably quite rightly) have to work hard to obtain and understand it properly.

Beyond supplier pricing there are a range of other costs that we could usefully target such as poor quality, packaging, inventory levels and obsolescence, tax, and customs duty. Section 4.7 may be useful here because we need to build a model of the total costs.

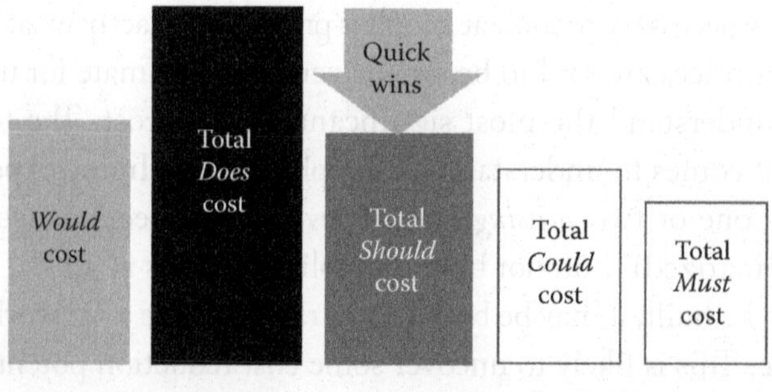

Figure 4.22 Stages of total cost reduction.

There are four distinct and useful stages to consider when carrying out cost reduction work, as shown in Figure 4.22. I learned and used these for many years consulting with Ingersoll Engineers/Bourton Group.

The total *Does* cost model is along the lines of total cost described in Section 4.7. It is very important to establish this and make this clear to all involved, otherwise, a new lower total cost will be compared with what people *thought* it *Would* cost. The idea of what it *Would* cost is sometimes out of date, may only include the price, or provide an incomplete cost model (for example, it may not include emergency shipments or premium service callouts).

The *Should* cost model is what the cost should be if the obvious measures were taken to reduce cost. You could call them the quick wins. They should not require redesign or significant new investment or change the functionality of the product. Common examples are changing the packaging, transportation, or removal of items that are supplied but never used by the customer. These cost reductions do not typically require a heavy burden of engineering or marketing reapproval and can therefore be quite quickly implemented.

Could cost levels are achieved by doing something more radical in terms of design, functionality change, resourcing to a lower labor cost area, or changing the product design or specification. It is likely you will require help from sales and marketing or engineering or manufacturing, and suppliers may also be a source of ideas and help. A cost teardown would also be a good approach, as described below.

Cost teardown examples are often featured in the purchasing press. They aim to be a thorough analysis and questioning of costs, for example, by modeling the total costs of a product. They typically go through the following steps:

1. Establish a breakdown of the total product cost into constituent elements. For a computer, this would detail the cost of components such as electronic chips. Their price and any other costs such as import tax and cost of failures would also be listed. The cost of certain functionality (e.g., a camera on a laptop) would also be costed to enable its cost to be compared with the value delivered to the customer. Marketing is then able to make an informed decision about removing the feature from the specification. Using the example of a computer, the disassembled computer is often set out in a room with all the components and areas of functionality costed.
2. Make a thorough and informed analysis of all the components and functions to examine the following:
   a. Can we identify ways of providing the same functionality at lower cost?
   b. Is the cost of each function justified by the value (and price) to the end customer?

   This analysis is usually made by a cross-functional group (e.g., marketing, engineering, manufacturing, and supply chain). The cross-functional discussions are useful in providing immediate discussion and feedback on ideas. Sometimes we have included suppliers in these meetings with quite good results. This has included suppliers looking at all the components to see if they can supply them at lower cost.

3. Identify practical cost reductions and make an outline action plan for how they could be implemented, at what cost, and how long they would take.

It is useful and effective to make one person accountable (typically the product manager) for getting the proposals checked out in detail in a relevant department such as marketing, engineering, manufacturing, or supply chain. After this, another teardown meeting is held to approve, modify, or reject each of the proposals and for an implementation plan to be started.

Must cost levels are defined by marketing as the cost at which products must be produced, otherwise they will not sell. They are a firm target and can be used with suppliers as well as for internal cost control. Suppliers need to understand the imperative of achieving a certain price level, sometimes with future reductions. Otherwise, they (and possibly we) will not get the business. It may be useful to start with some general price enquiries, but making sure the supplier understands the must-cost nature of the business at an early stage will avoid wasting time and effort if the supplier will never be able to achieve them.

Total cost reduction can have a significant impact on profit, but be certain it is real. It must be based on the total of all attributable costs, assessed practically from supplier's supplier to the end customer.

* * *

4.14 Lead-Time Reduction

Lead times are sometimes even more important than price. Customers usually expect delivery in less than the total supply lead time. As a result, we spend time and money on forecasting and *sales and operations planning* processes to optimize inventory and customer service levels. Even then, we have stockouts on the one hand and obsolete

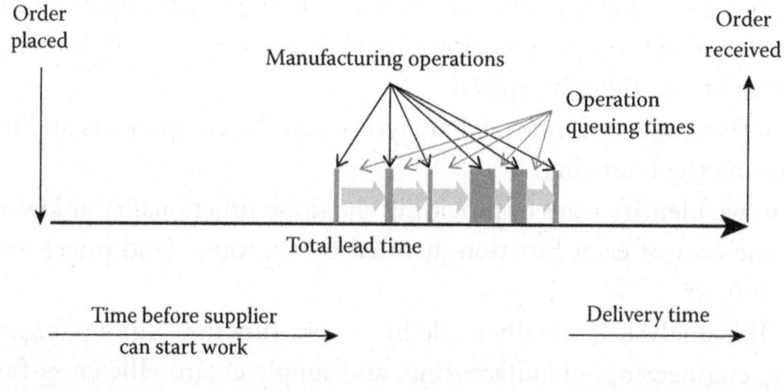

Figure 4.23 Lead-time reduction.

inventory on the other. Shorter lead times can reduce business costs significantly. In most new product introduction programs the need for short lead times is even more acute. Prototypes often need to be made as fast as possible because time to market is critical. Lead-time reduction can therefore be one of the most important supply chain activities.

Most lead times contain a small amount of processing or value-adding time, and a lot of queuing time, as shown in Figure 4.23. This applies within manufacturing processes and even more between suppliers of different stages of manufacturing. We sometimes do a lead-time efficiency study for a manufacturing process that divides the processing time by the total lead time. The results are usually around 1%–2%. The remainder represents a valuable target for lead-time reduction.

For repeat manufacturing processes, the reduction of queuing time before work can start is usually achieved by some form of capacity booking. For example, many businesses have an ongoing demand for a range of products but do not know exactly how many they will sell each month. The variation may be 50% each month. Commitments can be made to suppliers to take all the volume scheduled for (say) a 3-month period, but be able to flex the delivery date. This effectively shortens the lead time because we can keep rolling the schedule forward, committing to a new month's worth of volume each month.

Operational queueing times are under the control of the supplier or manufacturer. If that company has some form of cellular manufacturing, this time will usually reduce considerably. Regular orders (even if the quantities vary) can also help it reduce queueing. If quantities are not too variable, you may be able to implement a kanban system that pulls the product through each stage of manufacture. This can reduce the lead time to the time for the final operation plus the transportation time.

Getting a supplier to hold stock at a stage of manufacture is another approach. Clearly, the lowest cost stage that produces the most lead-time reduction is most advantageous. A good example is inventory of raw material. The arrangement can have a double benefit if the same raw material can be used to make more than one part or version of the part.

Long lead times on services such as equipment breakdown can also be very costly. Response times are a standard part of most maintenance contracts. However, sometimes these only achieve a first response. The need for a spare part is then discovered and a long wait ensues. A risk analysis (see Section 4.8) can be a good starting point. Mitigation against long breakdown times could include having critical spare parts in stock. The risk evaluation and mitigation should clearly be carried out with the key maintenance providers. This will add maintenance knowledge and experience, make it easier to work out practical mitigation actions, and increase the level of ownership from the supplier.

Sea freight time from Asia to Europe or the United States is a regular problem, and there is little to do about the 5 to 6 weeks lead time for which your inventory is tied up and inaccessible at sea. If customer service is important and the goods are not too heavy, some clients make a regular (therefore lower cost) arrangement to use indirect airfreight for a mix of product. They modify the mix of product they ship depending on their inventory level. For smaller and lighter shipments, this is not as expensive as you might think.

Effective lead time can also be reduced by adjusting planning parameters. When planning down through a BOM and applying the correct lead time at each stage, the total lead-time offset can be very long, in many cases in the order of 12 months. It can be instructional to set out a diagram of total product lead time. If you work down the BOM adding up the longest lead-time items at each level, the result will be a diagram of the total lead time and longest contributor items and stages. It is also useful to note the cost of the longest lead-time items. Low-cost, long lead-time parts or stages could be candidates for buffer inventory holdings. This sort of diagram can make it very clear where we should be concentrating on reducing the lead-time or holding inventory. The lead times on the diagram will be the ones used by MRP to suggest order placement. However, lead times are difficult to estimate accurately. They depend on estimates that we and suppliers make up, but they vary greatly depending on supplier inventories, workloads, and capacity. If we set lead times toward the worst case scenario, the resulting product lead times will be far too long. Resetting lead times on a regular basis is a time-consuming task and is usually not done. This is one of the reasons we prefer the use of MRP to drive outline schedules that are then pulled as deliveries by some kind of kanban or VMI arrangement. The effect of MRP parameters is discussed in Section 5.3.

The great majority of lead time is not value-adding. Reducing wasted time should reduce the need for inventory and improve customer service.

* * *

4.15 Contracts and Agreements

Supply chain contracts need to be clear and explicit about what commitments are being made, what is expected, what levels of performance are agreed on and measured, and what the remedies are for various possible problem scenarios.

We have found that contract discussions that start with the practical arrangements for doing business work better than those that start with the legal aspects. We often contrast our experiences making agreements with Western and Japanese businesses. The Japanese always need to make sure they understand practical arrangements—and in a sufficient level of detail. This often involves more than one run-though of each area. We have often been involved in these discussions in Tokyo hotels long after normal business hours have concluded: "Can we discuss the format for the monthly schedule update?" "What details do you need us to put on the electronic invoice?" By comparison, Western discussions tend to focus too much on prices and volumes,

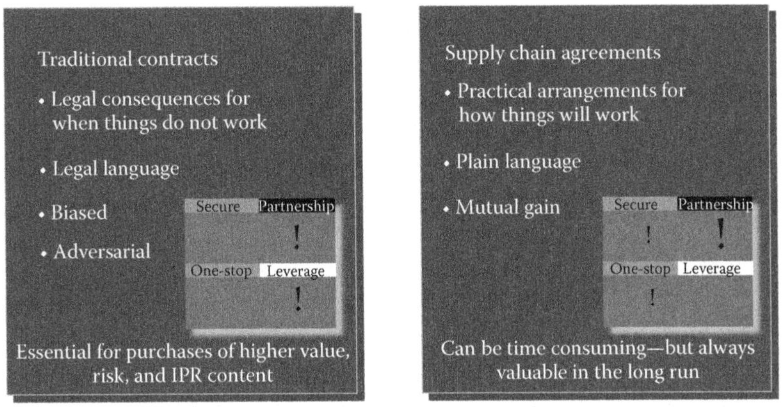

Figure 4.24 Contracts and supply chain agreements.

followed by use of a standard contract. On a two-day visit to a key new U.S. supplier we spent hours sorting out technical, logistic, volume, and commercial details, only for it to be suggested that "all we need to do is get my lawyer to talk to your lawyer and sort out the contract." We were not keen on passing the drafting of the contract to people who had not even attended the meetings. In the end we wrote the contract and the lawyers checked it. The only change they wanted was to base it on the laws of the State of California rather than English law. Our experience is that arrangements done the Japanese way do not very often go wrong. We are also amazed at how long standard contracts can be, even for very simple work.

We suggest you the use supply chain agreements rather than standard boilerplate contracts.

Figure 4.24 contrasts *traditional contracts* with *supply chain agreements.* We have written many *contracts* in the form described under *supply chain agreement.* That is to say, in plain English, covering the important practical arrangements and legal terms and conditions. There have also been many occasions where we have not been allowed to do that by the legal department and have had to make our supply chain agreement into an appendix to a weighty piece of boilerplate contract. We have also found that developing a supply chain agreement can provide contract writers with a valuable understanding of the supply situation that enables them to produce a better contract.

It is essential to keep a register of all contracts and NDAs or confidentiality agreements. The register should include *at a minimum*:

- Type/name of document
- Name of the other party
- Name of the person responsible (e.g., supplier manager)
- Date of expiry

One of the main functions of this register is to provide a way of planning and managing the maintenance of documents to enable them to be located quickly and to flag those that are close to expiring. This sounds like very basic business sense, but difficulty locating contracts is a common occurrence. Recently, a contract we needed to review was encrypted. Although the contract was available, the encryption key could not be located. The contract could not be read.

Ensure there is a clear understanding with the supplier about how business will be done in practice. Writing this down in simple English will help to clarify it and make a good basis for the contract. Keep a good register of all agreements—particularly when they expire!

* * *

4.16 Key Contract Clauses and Issues

The aim of this section is to provide an insight into the common contractual clauses, not to turn you into legal experts. In fact, do not do this job on your own; get legal help to scope out and finalize contracts. In general, I would always seek legal advice if the work is of very high value, impacts our business significantly, or includes high levels of IPR, or other secret information.

Note: The material contained in the following sections is for purposes of discussion and does not constitute legal advice or recommendations.

Each of the following headings uses a term that legal people recognize. There are then a few phrases and words typically used in the specific type of clause, followed by some notes and issues.

SCOPE OF SUPPLY

This Supply Agreement is entered into by and between … Supplier … and Purchaser … supply of titanium and steel products as specified in Exhibit A …

… apply to all purchase orders …

… does not constitute an order

… applies to all current and future divisions, subsidiaries

… in the event of a conflict between the provisions of this agreement and any other purchasing document used by either party, the provisions of this agreement shall prevail …

The scope of supply needs to be very clear. At the basic level, who does it apply to, what does it apply to, and where does it apply? This may sound simple, but can be particularly important in larger businesses, where there are different legal entities, and where businesses are trading globally or having a business relationship involving more than one type of product; we need to be specific. Where the contract includes the development of products or software we need to include it and particularly who will own the intellectual property developed. Where joint developments are included, who will bear what costs and contribute what resource. The scope of contracts is often for a type of part or service rather than one item. This is economical, even essential, but the whole range of purchases need to be thought through within the scope of supply.

Two other issues are usually covered. The contract is usually not the go-ahead to make or deliver goods. The scope wording usually states that this contract does not constitute an order. The contract provides the terms and conditions under which business will be done and the purchase order gives specific commitments for quantities, dates, and a pricing structure. Another of these clauses is referred to as precedence. It says that the contract terms and conditions cannot be overridden by any other document unless it is an approved and signed contract amendment. This is done to overcome the battle of the forms that used to be common. A supplier, for example, could override the PO terms and conditions by sending an order acknowledgment with his or her own terms and conditions (usually printed on the back using a font size that required a large magnifying glass). The battle of the forms could be continued by a document from the purchaser. The precedence clause prevents this.

CONFIDENTIALITY

.. supplier to hold … as private and confidential … not disclose without written consent … not apply to information … in the public domain … … independently and lawfully developed by …
may divulge to employees and Subcontractors on a "need-to-know basis"… take all steps to ensure obligations of confidentiality
… confidential information from the supplier … only be given in written form … marked "Confidential" …

Confidentiality must often be ensured even before a supplier is selected, let alone a supply contract signed. Discussions about technical and other issues often need to take place even to test the viability of what we plan to do. Confidentiality is important at this stage to prevent others from copying our idea. A nondisclosure agreement (NDA) is one of the first jobs to do before starting to even discuss ideas with another party.

Confidentiality or nondisclosure agreements can be with a business or individual. Suppliers who sign NDAs should also get their employees to sign a confidentiality agreement, otherwise, employees may be able to leave and use our intellectual property.

Confidentiality agreements are usually seen as protecting our technical ideas; however, it may also be important to cover other areas such as marketing.

The confidentiality of people information is also very important. More personal data is stored than ever before and disclosing it can be very damaging for the person involved and also costly if litigation against us results. In practice, personal information can be given to many suppliers who are providing people-related services (e.g., relocation, medical insurance, pensions). If you are involved in these areas, consult your legal advisor, make sure that you and your suppliers understand the rules, and stick to them accurately!

ENTIRE AGREEMENT

This contract embodies the entire agreement between the Parties with respect to the subject matter hereof, and prevails over any previous oral or written understandings, commitments, or agreements pertaining to the subject.

This clause makes life in court much simpler because it means that other documents, such as drawings and letters, cannot be argued in court to be part of the contractual agreement. The downside for supply chain people is that documents we want to be part of the agreement may be excluded. We therefore need to include such important documents as an appendix to the contract. Our contracts people will also have the opportunity to examine it.

NOTICES

Any notice or communication ... by hand; by registered or recorded delivery post; facsimile or electronic mail confirmed by first class or recorded delivery post within 24 hours of transmission.

This is useful to have when in dispute. It specifies how and to whom contract notices have to be delivered in order to be valid. The "whom" is usually a person's business title, not a name—for obvious reasons.

JURISDICTION

The construction, performance and validity of the Contract shall in all respects be governed by English Law ... State of Texas, French Law ...

The laws of many countries and states are similar and mostly quite fair. My opinion is that jurisdiction needs to reflect where the obligations of the contract will be carried out. For example, if you take an issue relating to goods being damaged during transportation in China to a court in Texas, they will probably not hear the case.

DISPUTE RESOLUTION, ARBITRATION, AND MEDIATION

If any dispute arises ... excluding matters referred by the contract for Expert Determination ... 7 days' notice ... center for Dispute Resolution.

Disputes can be taken to court, which is usually costly, takes time, and is not usually secret. Arbitration is being specified in contracts quite often either in final settlement or in the first instance. The contract specifies who will be the arbitrator; for example, the International Chamber of Commerce. Expert bodies are also used sometimes, particularly for specialist areas such as petroleum or software. On a recent train journey from London to Paris, I sat next to a lady on her way to the International Chamber of Commerce in Paris. She was going to present her case in a dispute with an organization owned by a foreign government where it was felt she would not get a completely fair hearing. Arbitration had been agreed to in the contract. However, arbitration is now becoming more expensive and mediation services are being used instead.

CONTRACT TERMINATION

... either party ... become insolvent or bankrupt ...

... shall have the right by prior notice to enter that others premises for the sole purpose of removing any item which are its property ... clearly marked as such... full and unfettered access ...

... For the purpose of the Contract the term Force Majeure shall mean:

war and other hostilities ...

riot, commotion, or disorder ...

earthquake, flood, fire, or other natural disaster

Neither party shall be considered to be in default of its obligations to the extent it can establish.. Prevented by ... Force Majeure

notice of contract termination may be given on 60 days' notice by either party without cause ...

... either party ... become insolvent or bankrupt ...

... shall have the right by prior notice to enter that others premises for the sole purpose of removing any item which are its property ... clearly marked as such ... full and unfettered access ...

... For the purpose of the Contract the term Force Majeure shall mean:

war and other hostilities ...

riot, commotion, or disorder ...

earthquake, flood, fire, or other natural disaster

Neither party shall be considered to be in default of its obligations to the extent it can establish.. Prevented by ... Force Majeure

notice of contract termination may be given on 60 days' notice by either party without cause ...

Contracts can be terminated in a variety of circumstances. How they can be terminated and what happens afterward is a very important contractual subject—and one that is often in dispute. A contract is usually for a specific period of time, a specific quantity of goods, or specific tasks. Some contracts specify a period of time, but allow for renewal under specific conditions, and some are even evergreen. Contracts usually have specific terms for termination and most specify breaches of contract that will result in termination. The arrangements for termination, renewal, and what happens after termination need to be clear.

Most contracts have requirements which, if not met, will result in contract termination. In these cases, where the supplier is in default, he or she may need to pay a financial penalty or may not be paid at all. The *force majeure* clause specifies where parties *will not* be in default because it was caused by circumstances beyond their control (so-called acts of God). Legal people have practiced and modified these clauses over the years to reflect how they work in practice. They are carefully worded and my advice is, do not change them without very good reason and get legal help—do not change them yourself!

The contract also needs to cover what happens after the contract has terminated—who owns the tools, equipment, or software used to make the product? How will services or spares be provided, and for how long? Who owns the intellectual property rights? This is particularly important if the contract is terminated because one of the parties is in default.

Access to goods we own but are in the possession of a supplier who is declared bankrupt or is under administration is also covered. These clauses often require a court order to enact, which takes time. A visit to the supplier site, taking a clearly worded contract (unfettered access) and proof of ownership can sometimes secure the goods.

INTELLECTUAL PROPERTY RIGHTS

All intellectual property rights... under this Contract which are written or produced on a bespoke or customized basis ... future such rights when the said works are created ... be owned by the Purchaser ... where the Supplier provides existing intellectual property right protected material to the Purchaser under this Agreement it shall disclose this to the Purchaser.

Interestingly, even when a customer has paid a supplier to develop some intellectual property, it will usually belong to the supplier unless the contract specifically states otherwise. If we want the intellectual property to belong to us (the Buyer) the contract needs to make this clear. This can be applied to designs of components or products, software, or research.

Copyright also needs to be protected and applies to circuit board layouts, documentation, marketing materials, and photographs. Without specific contract terms, the creator of the materials automatically retains certain rights, even if paid by the purchaser for the work.

Trade secrets are often protected by contract in the same way. These can be knowledge about how a process works or what temperatures or chemical mix has to be used. Suppliers may learn this kind of information as part of working for the customer.

TERMS OF PAYMENT

Provided that the invoice ... was entitled to submit at the times and in the manner of the contract ... shall pay within 45 days ...

If the (Purchaser) shall fail to make any payment to the (Supplier) to which the (Supplier) is entitled under this Clause ... then the (Purchaser) ... shall pay ... in addition to the amount not properly paid simple interest for the period until such payment is made. The rate of interest shall be ...

Clear terms of payment are usually included in the contract. Penalties for late payment are often included, but in many countries are rarely exercised. The problem is that getting the payment may be made more complicated by the introduction of another request for payment. There is some European movement, driven at government level to reduce late payment, particularly of small businesses by large ones. Scandinavian countries provide a good example of timely payment. It is expected that customers will pay on time or they will be taken to court quite quickly. Judgment will be quite fast and penalties will be imposed.

Discounts for early or on-time payments are common—1% for payment within 30 days, for example. Suppliers find that these terms are more easily administered. However, some large company payment processes are just too cumbersome to pay that quickly.

TIME IS OF THE ESSENCE

The time for delivery and/or completion of the work to be performed under the contract shall be the essence of the contract.

Try dropping in the *time is of the essence* phrase in a gathering of legal types and you will probably find heads turn to check what you are saying. This is because liquidated damage clauses are likely to be linked to this clause and may result in large compensation payments.

Time is of the essence clauses cannot be used to improve delivery-on-time performance. They are for use in cases where a supplier's failure to deliver or perform a service before a specified time period causes the product or project to be in danger of failing. In these cases the liquidated damages will reflect the value of the complete project rather than that of the goods or services of the supplier. An example could be the supply of fuel oil to an oil and gas exploratory facility in the North of Alaska or Russia. Failure to transport the fuel to the camp before the ice roads have melted will mean trucks cannot make the journey. This may result in loss of work and monetary sums far in excess of the value of fuel or the transportation costs. The liquidated damages clauses would bring these higher values into play. When entering into these contracts, suppliers usually need to take out insurance against failure and the exercising of liquidated damages.

ASSIGNMENT AND SUBCONTRACTING

Neither party can assign the contract or any of its rights ... without the written approval ...
... supplier will be responsible for the acts, defaults, and omissions of its subcontractors ...

We may choose to prevent the supplier from subcontracting all or some of the work he or she is doing for us. Where the supplier is allowed to subcontract, we should specify what he or she can and cannot do and also specify which areas of their contract with us must also apply to their subcontractors. For example, in one case a company had contracted for transportation of equipment over a long distance. One day, they followed the supplier's vehicle from his premises, only to see it enter another yard. His goods were then reloaded to an inferior vehicle for transport to the destination.

Trade compliance and control (TCC) is mentioned here because we need to be aware where our work is being carried out and ensure that TCC rules are not broken. One example is of a company who contracted with a fourth-party logistics provider (4PL) to transport goods from the United Arab Emirates to Kazakhstan. The 4PL contracted an airfreight company to move the goods. They later became aware that the aircraft with their goods on board would stop to refuel in Tehran. The goods in question contained U.S.-controlled content and even a brief stop in Iran would be a breach of U.S. TCC rules. The goods were transferred to another aircraft.

STATUTORY REGULATIONS

Both parties … agree to comply with all Laws … all applicable export regulations … regulations … cost borne by that party
If the cost … is increased by … after the date of the contract … Any new law … regulation … amount … paid to the supplier
… undertakes that he, his employees and subcontractors will … comply with all health and safety requirements …

Some of the important areas included here are health and safety, environmental regulations, import/export, and trade regulation rules. There are some interesting legal consequences here—for example, if a supplier or service provider is subject to environmental regulations that are changed by a particular country. If the new regulations result in them bearing higher costs, the laws in most countries will enable them to pass the cost increase on to their customer.

WAIVER OR THE NO-WAIVER CLAUSE

Any failure by the purchaser to insist upon the performance of any of the terms … not constitute a relinquishment of … shall continue in full force.

The idea of this clause is to ensure that if we do not raise a complaint over a supplier failure (e.g., a part defect) perhaps even over a long period of time, a subsequent complaint about the same failure will still be valid. If we do not have such as clause in the contract, the supplier might argue successfully in court that the fault cannot be important because we did not complain before.

INCO TERMS

- DDP—Delivered Duty Paid—Maximum responsibility on Seller for delivery and assuming duties and risks.
- EXW—Ex Works—buyer bears all responsibility for collecting goods.

Inco terms are an excellent way of specifying exactly when responsibility for goods is passed from supplier to customer in a highly effective way, using only three or four letters. This means that after this point, the buyer is responsible for all ongoing costs, including insurance. Some definitions I have read state that this is the point where ownership transfers from the seller to the buyer. In my experience, this transfer of title need not take place at this point. Many will defer this to the point at which the goods are paid for.

Shipping methods are changing and containerization is replacing the old methods of loading the product onto cargo ships with cranes. Free on board (FOB) is therefore being used less, but I particularly like the definition by the International Chamber of Commerce, which states that transfer is made when "the centre of the hook of the crane traverses the guard rail of the vessel." The need to be exact is presumably because in the past, some goods have fallen between the dock and the vessel and a dispute has arisen. This might be important if the goods are chemicals that have then leaked into the estuary and poisoned all the fish and would result in claims from fishermen for loss of livelihood.

INTELLECTUAL PROPERTY WARRANTY

Supplier shall.. Indemnify and hold harmless Buyer ... from and against ... all actions ... of any actual or alleged infringement of any patent copyright, trademark, trade secret ... except to the extent such infringement results solely from the manufacture of such Items pursuant to detailed designs furnished by the Buyer.

This applies to software or other products which may contain intellectual property (IP) that may belong to someone else. This type of clause is to help to protect the buyer when a supplier provides a product that contains someone else's IP. The seller holds the buyer harmless from the actions of those who own the IP. In order to hold the buyer harmless, the seller is usually required to insure himself or herself against such actions.

LIMITATION OF LIABILITY

In no event shall Buyer be liable to Seller for anticipated profits or incidental or punitive or consequential damages. Seller hereby waives all rights to such damages.

These clauses usually limit the liability of the *buyer* to the *seller.*

EXCLUSIVITY

... shall maintain said resources solely and exclusively for the use of (Purchaser) ...

It would sometimes seem to be a good idea to contract for exclusive use of a facility or some machines, for example. We may be worried that we need all the capacity of the equipment, or that our competitors may also use the facility and be able to view some confidential work or design.

The problem with doing this is that if we do not use the facilities to the level of capacity we agreed to, then we are still likely to be responsible for paying for them. It acts almost like a take or pay clause.

It is important that you think through the arrangements that you need to be covered by the contract and explain these to your contracts advisor or lawyer. If you do not do this, the contract will be a standard one that may not cover important special risks and issues.

* * *

4.17 Trade Regulations and Compliance

Trade regulations, and in particular the U.S. Trade Control and Compliance (TCC) rules, have become a serious worry to many businesses. I say "worry" because it is so difficult to keep up with these changing regulations and ensure compliance as political positions and technologies change. Ensuring compliance requires most of the larger international businesses to have experts on their staff.

The internationality of trade and politics has become quite complex. End users are often uncertain where their purchases have been manufactured. European and U.S. manufacturers have outsourced a lot or sometimes all of their manufacturing to Asia. Some of the Asian product includes designs or components manufactured in Europe or the United States. Some of the components carry restrictions on where they can be exported to. For example, I recently bought a new computer product of U.S. design (but not manufacture) using a website in the United Kingdom. I had to certify where I would take and use the product. This is because U.S. trade regulations last for the lifetime of the product and apply after the original manufacturer has sold it.

There are specialists in law and trade regulations, but in the front line here, sourcing the product on a global basis, are supply chain people. The regulations are driven by fast changing political events. For example, at the time of writing, Russia has been subjected to trade restrictions because of the political situation regarding Ukraine. Hopefully, these will soon be resolved. It is unrealistic for all supply chain people doing global work to be fully conversant with the latest trade regulations. It is more realistic to have good background knowledge and to rely on a specialist when required. Certainly, if we were able to give all the current details in the following pages, it would be out of date before we printed it.

It is important for supply chain people involved in global sourcing or transportation to know enough background information about trade regulations to know where it will affect longer-term sourcing and when to gain expert advice.

* * *

Many countries have applied their own important restrictions for some goods and technology, mainly when the use of these items can be applied for other (dual-use) military activities. However, the Wassenaar Arrangement involves 40 countries and is based on a common agreement to regulate control products as per "dual-use" items. The Russian Federation is a member of the Wassenaar Arrangement; this means that the Russian Federation controls many of the same goods as the United States and Europe and export licenses are required. Both the United Nations and European Union also have specific rules. These are in addition to specific rules made by countries; for example, Norway, Sweden, and Singapore all have their own rules.

The most stringent regulations are those of the United States, which cover a wide spectrum of trade:

- All transactions of products, services, and technology that are manufactured in the United States or shipped from the United States (regardless the origin)
- Foreign items that contain at least 10% U.S. content
- Foreign items using controlled U.S. technology
- Deemed exports (technology transfers)
- Comingled software (U.S. and non-U.S.)

U.S. trade regulations are unilateral and follow U.S. government interest, but are also in place with extraterritorial application.

U.S. embargoes exist on a range of countries which, due to changing political issues, changes quite regularly. There are currently U.S. trade regulations on many other countries such as Iran, Sudan, Syria, Myanmar (Burma), and North Korea. These regulations (and the countries they cover) also change regularly. The oldest rule is the Cuba Embargo, where most transactions are limited and prohibited. At the time of writing it seems that the relationship between the United States and Cuba is improving and we may see an end to the embargo on Cuba in the near future.

Under the U.S. regulations

1. No U.S. person may be involved
   a. No facilitating operations, no providing technical assistance, no communications on commercial matters, no business planning, no advice on any transactions, no approvals

      A U.S. person is any person who is a U.S. citizen, holds a green card, or is a resident in the U.S.

2. Non-U.S. persons may
   a. Export goods, services, and technology to Iran unless they are controlled by local laws or by reexport regulations (U.S.)
3. The only U.S.-origin items allowed for reexport are
   a. Any items incorporating less than 10% U.S. content (*de minimis* calculation)
   b. Iran and Sudan: EAR99 items in inventory (Export Administration Regulations)
   c. Items approved on U.S. license exception—TCC involvement essential
4. No direct exports or transshipments to Cuba, Iran, or Sudan from the United States

Many countries and organizations have trade regulations. Those of the United States are the most far-reaching and comprehensive. If goods have U.S. content, have been to the U.S. or have U.S. patents, be very careful—there are likely to be restrictions on their use. U.S. persons are also covered by these regulations in terms of what and with whom they can be involved.

* * *

The export checklist below lists the issues that export control will need to be satisfied regarding all jurisdictions:

1. Country of origin (COO):
   a. Where is my product manufactured?
2. Classification of the product export control classification number (ECCN):
   a. Dual-use items/military controls
3. Country of final destination:
   a. Embargoed destinations/terrorist-designated countries
4. End user:
   a. Government/prohibited parties
5. End use:
   a. Nuclear end users/chemical weapons

Breaking U.S. trade regulations can result in huge sanctions such as the revoking of U.S. export licenses for 20 years. This would effectively mean many businesses would have to cease trading.

Be very careful out there. For example, note U.S. trade regulations say that trans shipments via Cuba, Sudan, or Iran are against regulations. This means, for example, that an aircraft carrying your goods cannot stop to refuel in Iran.

* * *

4.18 Bribery and Corruption

There has also been a drive against bribery and corruption, particularly by countries such as the United Kingdom and the United States. In the United Kingdom, the

Bribery Act was introduced under which criminal prosecutions are brought. The U.S. Federal Corrupt Practices Act (FCPA) is more wide-ranging and can result in very large fines. Bribery can take many forms; a few examples are

- Giving or taking gifts or money in order to win business
- Giving or taking gifts or money in order to facilitate a process such as importation
- Giving or taking money to gain permission or approvals for facilities (e.g., to build a factory)

Bribery is sometimes difficult to define or at least to know where to draw the line. For example, is it a crime to invite some of your customers to a sporting event and to an evening dinner afterward? Probably not. Is it illegal if you also pay for them to stay in a hotel that evening or even pay for their flight from their home country in Europe? Getting tricky, but probably not. Would it be bribery and against the Act to also pay for a man's wife to accompany him, and pay for her airfare and hotel? This is probably beyond the limit and is against the Act. The U.K. Bribery act will build up case law as a guide to where to draw the line.

Large fines have been levied and paid under the FCPA. These include incidents that have occurred outside the United States and to parties who are not U.S. persons or businesses. These have included fines paid by shippers on behalf of their client to expedite importation procedures in certain countries. In these cases, both the shippers and their clients have been fined amounts of many millions of dollars.

As a member of a supply chain organization you are likely to find yourself doing business in a foreign country. Bribery may be an expected way of doing business in some of these countries, but it is not a good idea to go along with it. You will somehow be caught in the end, but in any case, it is wrong and perpetuating such systems is the wrong thing to do. A good approach is to make it clear that your company has very strong ethics and rules and will not accept this type of behavior. It has been commented about some of our clients that suppliers are happy to know where they stand following this sort of clarification.

Regarding the difficulty of deciding where the limit of acceptability versus unacceptable is, you will find it useful to think these through beforehand rather than have to make a snap judgment. Will you accept any gift at all? Will you go to a sporting occasion with a supplier? Will you ever play golf or go to dinner with a supplier, and would you go only on the premise that you paid 50% of the time? You need to make these rules for yourself. They are certainly influenced by where you are in the world. Dinner with a supplier may not be all bad—especially if one of you is in a foreign land. You may gain an understanding of the supplier or salesperson that you would otherwise miss. I certainly do go to dinner with suppliers sometimes, but I do make it clear that I will pay my way.

One employee of a client (we have a deep admiration for this individual) has worked in a range of overseas posts and been in a number of difficult situations regarding ethics. One of these involved obtaining a fire certificate for a newly built factory in a foreign land. The local fire chief informed him this was a simple matter of providing him with (the equivalent of about $250) cash and the certificate would be issued. However, our friend wanted the factory's fire facilities to be checked by professionals to ensure it was properly safe. Even so, the cash was expected personally by the fire chief. After much discussion involving facilitation by another local official, arrangements were made for no bribe to be paid, but for the local fire station showers to be upgraded at the cost of the company. The new shower block now carries the name of the company as its "sponsor."

Establish your rules about what you will and will not do. Make clear your own rules and those of your business to any supplier or business partner.

* * *

4.19 Corporate Responsibility

There have been a number of high-profile cases recently where European retailers have been damaged because of unscrupulous business practice further back in their supply chain. Some of these have involved child labor and one of the worst involved deaths in Bangladesh when a subcontractor's building collapsed. Should we be responsible for the practices of businesses who are suppliers or subcontractors to our own suppliers—businesses we may not even know about and have certainly never visited? These cases may be difficult to apply specific laws or regulations to, but if we are a business outsourcing manufacture, the answer is of course yes.

I was part of a factory visit in Asia together with a client and a representative of their owner, a PE company. We became concerned about a number of practices related to the health and safety of employees in this factory, which included guarding on some of the machinery and fire precautions. Even though the business was not supplying to our client directly (they were a supplier's supplier), we were obliged to give notice to the supplier that unless our concerns were addressed using an action list we drew up, the business would be withdrawn. The representative of the PE company was particularly firm and very clear in his explanation of their ethics policies.

It is time-consuming and sometimes expensive to visit suppliers who are a long way away. However, in my experience this is absolutely essential for a number of reasons:

- You will only be able to see what really happens by visiting the supplier factory
- You will be able to understand much more about the supplier, his or her manufacturing methods, process quality, and control
- You will develop a better relationship with the supplier

Sometimes, you will be faced with circumstances that require personal strength of mind and standards. Some discussion of this is contained in the ethics section. It is too easy to accept the assertion that *this is the way we do things here* or *this adheres to local standards.* When we know something is wrong or dangerous we need to speak up and insist that things are put right.

Corporate responsibility is becoming more important as we outsource to distant places. We are responsible for the standards of our suppliers and their suppliers. Supply chain people are in the front line when it comes to spotting the issues and resolving them.

* * *

4.20 Regulations Regarding Electronics and Substances

There are regulations about the use or transportation of a wide range of goods. Live animals, pharmaceuticals, flammable goods, batteries, and chemicals are typical examples. These certainly affect the supply chain people responsible for buying, moving, or storing these goods. Supply chain people certainly need to know about the regulations, use experts where necessary, and make sure requirements regarding sloping warehouse floors, drainage, and unsafe colocation of chemicals are implemented.

Knowledge about some of the regulations is essential for product design programs. For example, lithium-ion batteries are classed as dangerous goods. Depending on their type, size, and shipment as part of another product, they are subject to various regulations. These rules need to be known by product designers. It is far too late if it is only when you try to ship prototypes somewhere that you are asked to sign a dangerous goods contract by your shipper!

The Waste and Electronic Equipment (WEE) Directive was introduced in the United Kingdom in 2006 and was replaced by new regulations in 2014. The new regulations are more specific and include more products. Further products are listed for the legislation to be applicable to from 2019. WEEE includes most products that have a plug or need a battery. There are 10 broad categories of WEEE currently outlined within the regulations. These include household appliances, computers, sound equipment, electrical tools and equipment, toys, medical devices, and so forth. The regulations place a range of responsibilities on businesses supplying this equipment related to identification of certain types of items and substances and their methods of disposal. This is another important area for the product design teams and the supply chain to know.

Regulations regarding the use, transportation, and disposal of various products are a key consideration during the product design process.

* * *

4.21 Auditing Requirements and Sarbanes–Oxley

When I had a line management job responsible for inventory, I needed to justify to auditors that my inventory valuation was done correctly. They would ask about the value of electronic items we were no longer selling and the level of resale value I had set against them. I had to justify the obsolescence provisions against inventory. This could be an alarming value and instructive for those trying to come up with a cost of holding inventory.

The Enron scandal in the United States reemphasized how a business that did not properly assess its value could lead to financial disaster, including investors losing their money. This in turn led to the Sarbanes–Oxley Act in 2002. This is a U.S. law and applies to businesses registered with the Stock Exchange Commission; however, other countries have increased the stringency of their own legislation in a similar way. For example, the United Kingdom introduced similar legislation in the Companies (Audit, Investigations and Community Enterprise) Act of 2004.

These regulations affect the supply chain where control procedures of the business are concerned. Bearing in mind the financial impact of the supply chain on most businesses, these include processes for buying, selling, keeping inventory records, and applying allowances for inventory value. The Sarbanes–Oxley Act (or SOX) has required new checks to be carried out by auditors and in the early implementation it was estimated that a large firm would see its audit fees rise significantly (e.g., from $250,000 to $750,000). Auditing of companies for SOX compliance or many other country legislations will require that control procedures be clearly specified and compliant. These procedures are basically aimed at ensuring that a company is properly valued—avoiding further company financial scandals.

Properly documented supply chain processes need to ensure purchase and sale transactions are done legally and that the true value of the business is represented by book values and allowances.

* * *

5 Logistics

The scope of logistics varies depending on the business. Here, we include planning of requirements, inventory management, and distribution to customers. Logistics people, processes, equipment, and systems are a large part of the order fulfillment process and of creating customer satisfaction. Getting the main processes—sales and operations planning, material requirements planning, inventory management, and distribution management—to work is essential. Most of the work requires energetic attention to detail and accuracy. The work is often complex by nature; however, some of the techniques included in this section are aimed at reducing complexity to a practical level.

5.1 S&OP

S&OP is discussed in more detail in Section 3.2.

This process of balancing supply to meet changing demand is one of the most important processes in many businesses. This should not be a tactical task left to junior staff. This is unfair to them and does not allow some of the important risk decisions to be taken that may enable excellent customer service without excessive risks in manufacturing capacity or supply chain disruption and expense. Figure 5.1 shows the concept of the process and is similar to those often used by the late Oliver Wight in his many excellent books on the subject. One of his books I still find useful is *MRP II: Unlocking America's Productivity Challenge*. This book was published back in 1984!

The process usually cycles on a monthly basis according to a time scale managed by a master scheduler, chief planner, or the S&OP manager. At the beginning of the process, sales and marketing present an updated forecast and data about the outstanding customer orders. It is important to synchronize the forecast with actual orders so that there is no double-counting or demand missed. When customer orders are large in value, it is practical to check them against prospects in the forecast. For higher-volume and smaller-value customer orders only cutoff points, larger orders, and trends can usually be checked. The forecast is more useful if, rather than simple numbers, it is accompanied by information about the risks and potential level of variation that could be expected. For example, elements of a forecast for a product that include demand from a potential new customer where negotiations are in progress should be highlighted and understood. There will be a risk that the new customer will not be won,

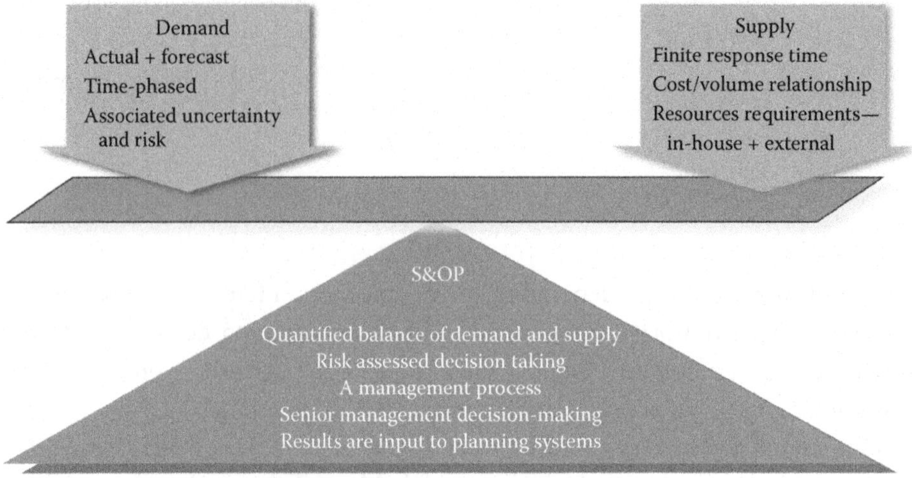

Figure 5.1 Sales and Operations Planning (S&OP).

but conversely, if the business is won, the customer may expect fast delivery for which we need to make provision in our own and in our supplier's capacity. This is exactly the type of risk decision that senior people will consider.

The S&OP manager will assess the new levels of actual and forecast demand and how current finished inventory, manufacturing, and procurement plans match. There are likely to be areas of shortage and areas of oversupply that the S&OP manager ideally should rebalance. Pushing out manufacturing plans is likely to cause inefficiencies and supply schedules might already be within the *firm window* or about to be delivered. The S&OP manager may decide to buy or build to the existing plan and increase finished stock if he or she can see that future demand will soon reduce the inventory again. If the S&OP manager cannot see finished inventory reducing, he or she may decide to recommend immediate reductions—at a cost. Pulling in or increasing build and supply schedules is also often required when demand increases. If this is done without assessing that capacity is available and suppliers are capable of delivering additional supplies, the plan is likely to be unrealistic and unachievable. This will result in customer service deterioration and additional costs in manufacturing, the supply chain, and inventory.

The S&OP manager will present a recommended new production plan to the S&OP meeting together with information about the forecast uncertainties and additional costs that he or she used to make recommendations.

The S&OP meeting typically includes the heads of the following functions: sales and marketing, operations, supply chain, finance, and the CEO (usually the chairperson). The S&OP manager (or master scheduler) usually circulates his or her recommendations and supporting data and issues before the meeting, attends the meeting to present the new plan, and should record any agreed adjustments to the plan and circulate the results with summary minutes.

Following the S&OP meeting, the S&OP manager is responsible for communicating to the following people about changes to the plan:

- Supply chain: To enact changes to supply schedules
- Operations: To enact changes to operations plans and capacity
- Sales and marketing: To enable realistic commitments to customers to be made

The approved plan is typically used as an input to MRP, which is dealt within the next section.

Most ERP systems include some form of S&OP function and these range from very basic to quite complex and from the standard to entirely customizable. Spreadsheet-based systems are used by a significant number of businesses, particularly where there are unusual requirements or the business is small or medium in size. These systems are generally criticized, usually for the same reasons most spreadsheet systems are

- They need to be interfaced with the other (usually main package) system
- They can have separate and out-of-date data or require significant additional data maintenance
- They need to be backed up separately
- They tend to be in the domain of one user and are difficult to keep secure

However, despite these criticisms, we have found a number of very good spreadsheet-based S&OP facilities in operation. They do seem to be able to produce a low cost and bespoke system for many businesses, and if the shortcomings mentioned can be addressed, we see them as viable solutions.

Sales and operations planning is one of the major business processes and is important to senior management for managing the balance between supply and demand.

* * *

5.2 Rough-Cut and Finite Capacity Planning/Scheduling

Finite capacity planning analyzes at the load on each significant work center and checks if it exceeds capacity. If it does, the work is rescheduled to avoid overloads. This description oversimplifies the work of this function, but in essence this is what it does.

A typical process sequence would be for S&OP to set overall production volumes and service levels, MRP to plan specific supplier and internal work requirements, and finite capacity planning to check the load against capacity for significant work centers. Depending on the IT facilities, MRP and finite capacity planning can run almost continuously.

The methods for rescheduling work-center load vary considerably. Work can be brought forward, transferred to alternative work centers, or pushed out to a later date. If work is pushed out, it may impact customer orders. Stand-alone *finite capacity planning* packages pass new proposed dates back to the MRP system to work out the consequences. In many modern ERP packages, the two functions are integrated and scheduling implications can be worked out within the one package.

My experience of finite capacity planning is that it is fairly complex. For example, it is likely that an overload in one area could require replanning of a sequence of operations. This may or may not affect the customer service level.

My experience in cellular manufacturing suggests the work of detailed scheduling is better done within the cell rather than in a more central function where real details of the situation are not known. My preferred way of working with cells is to use *rough-cut capacity planning* to check that the loads on key work-centers is reasonable. A rough-cut analysis of capacity should be made at the S&OP stage. (This is at least as important if you're using finite capacity planning because it should prevent you from overloading work centers too much.) After running MRP, orders and loads can be assessed by individual cell and appropriate action taken. At the cell level, more detailed information will be available than any central system can sensibly carry. Shifts, holiday arrangements, and alternative machines and methods can be used to achieve schedules. Cells may be working to specific work orders, or the MRP output can be treated as a guide to what will be required and kanban arrangements used for detailed control.

The choice of how to *architecture* your planning processes will depend on

- The complexity of your supply and make process
- How any internal manufacturing processes are set up (cells, flow line, etc.)
- Your IT package

The choices require thinking about a range of organizational issues as well as planning and IT systems.

5.3 MRP

MRP in its simplest form is a planning tool to calculate, from a top-level plan, what parts and resources will be required over a period of time. Input data required is shown in Figure 5.2. The figure shows the calculations for a kit of parts 50092 for which there is a revised master production schedule (MPS) or plan for production or usage. There is also a BOM showing all the parts and materials needed to make the kit. The example is for only one of the parts in the BOM, part 53291P1. MRP sets out calculations over an appropriate time period, showing the quantity of the part needed for the kit. In this case, the BOM shows two are needed for each kit and therefore 40 are required in period 4, 60 in period 9, and another 40 in period 13. Notice the requirements are 1 week earlier than the kit build because MRP has a 1-week lead time for the kit. Note also that there is no requirement for components for the kits to be made in week 1 because the parts have already been issued against a work order that MRP knows about (see work orders Section 5.3.3 and in Figure 6.4, Section 6.2 Supply Chain Information Technology). In a real MRP run, the demand for 53291P1 from all other parent kits or products would be included in this line. This is called

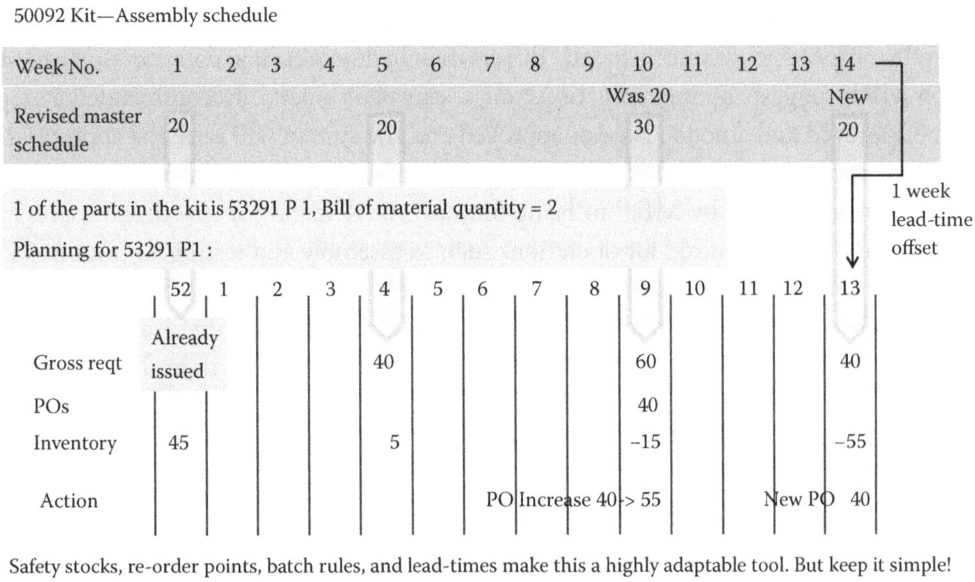

Figure 5.2 How MRP calculations work.

the *gross requirement or demand.* From the gross requirement, the MRP calculation progresses along in a time-phased manner deducting the gross requirement from the opening inventory and adding on any receipts due in order to calculate the *projected inventory level.* In Figure 5.2, the projected inventory is only five after the kit requirement in period 4 and is negative in period 9, even though a planned receipt of 40 has been taken into account. In period 13 the kit demand for 40 parts causes the projected inventory to fall even further down to –55.

MRP then decides what action should be taken, and in the example suggests that the PO for 40 expected in period 4 should be increased to 55 and a new order for 40 should be raised. In the example, 52391P1 is a purchased part, but if it was a part manufactured from raw material or from constituent parts, MRP would continue to plan down to the next level in the BOM, using this net demand as the basis for their requirements. This example is very simple and MRP could potentially plan the component quite differently based on planning parameters set for each part, for groups of parts, or for the way MRP works in a particular business.

Planning parameters vary slightly depending on the ERP package being used. A range of the most important and most popular are discussed below. In this example, the orders suggested by MRP assume orders can be raised and can be delivered on time. This is why MRP is often referred to as a *can do* system. There are parameters that prevent this to some extent, but Oli Wight invented the term MRP2 for an MRP process that cycled around the calculations the first time, checked that the actions suggested were *doable,* and adjusted the plan where they were not.

MRP output is generally routed to buyers or schedulers for them to arrange with suppliers. Parts are coded to buyer groups to enable the correct part planning results to be sent to the right people. Output used to be printed in the form of buy sheets, but generally output is now screen-based. Buyers or schedulers review data for all parts on which MRP suggests a range of actions: raise, cancel or amend. Recommended action can be approved and amended or not approved and the system will raise the appropriate supplier communication. This is usually some form of electronic data interchange (EDI).

Lead times are used by MRP to bring forward the demand for constituent parts by the amount of time required for operations such as assembly and testing. In Figure 5.2, the kit has set with a 1-week lead time. All the gross requirements for the constituent parts have been brought forward 1 week. Many packages offer a further lead-time offset capability to allow some parts to be requested either earlier or later than the other constituents. So if the general lead time for 50092 is 2 weeks, but the constituent 53291P1 has an offset of 2 weeks for its use on this kit, the part is deemed to be used at the same time as the kit is required (i.e., there is no assembly process time for this item on this kit).

Safety stocks are widely used to reserve some parts or material in the calculation at all times. If a safety stock of 20 had been set for 52391P1 in the example above, an order (for at least 15) would have been suggested for period 4. Subsequent order suggestions would also have been modified accordingly. Some systems will not raise orders within lead time if they are to meet the requirements of safety stock—they will raise the order to be due in the first week beyond the lead time.

Order-raising rules generally determine what size of order should be raised.

A *fixed lot* size parameter simply raises orders of the quantity specified. This is a blunt instrument, but it has its uses if, for example, an entire piece of material has to be used in one go or a full container load has to be ordered. If the demand is only slightly higher than the fixed order quantity, two orders will be raised, which may be many more than required. This parameter can also be a problem when demand for the part falls to much lower levels (e.g., toward the end of its life) and orders that are far too large may be raised.

Minimum lot size is slightly more sophisticated. An order for at least that quantity will be raised—and more if required. This is useful if, for example, we have agreed to a minimum order quantity with a supplier.

Lot-size multiplier parameters can be used where parts are generally supplied in containers of a certain quantity, such as boxes of fasteners, tubes of integrated circuits (ICs), lengths, and weights.

Forward period order cover (FPOC) is a popular parameter because it will suggest POs to cover a specified period of time against the predicted requirement (e.g., the next 6 weeks). This parameter, like many of them, can be used in combination with others, so you can usually use FPOC in conjunction with a lot multiplier.

MRP planning output can be much more usable if parameters are used to drive the order calculating algorithms to suggest orders that mimic the quantities and dates we would ideally raise ourselves. We do personally have the emotional scars resulting from overcomplex use of parameters that led to ordering mistakes. Most of the really big problems come along when there is a change in demand, timing, lead time, product specification (engineering change), or anything that can cause a significant change to demand or supply. For example, we could be running with an item used as part of a product, but with significant additional demand from spares and upgrades. If the upgrade demand suddenly dips severely and unexpectedly and the planning is not amended, very large-scale overcommitments can result.

5.3.1 Planning, Scheduling, and Pulling

We like to emphasize that MRP stands for material requirements *planning*. Our preferred way of operating is to use MRP to provide suppliers with our best information about our demand for their product and then use another tool to pull the final deliveries to us on a regular basis that is much nearer our real demand (i.e., it can react to our own tactical variations in demand). Figure 5.3 shows the sort of arrangement we like. A long (12 months in this case) schedule is provided to suppliers, in this case in weekly *buckets*. In the short term, deliveries are all controlled by a kanban pull system or a VMI system. Figure 5.3 shows how this can work in MRP and in conjunction with pull systems and the contractual terms and conditions that include allowance for flexibility in the schedule.

In this case, supplier orders for the first six months are committed in that the total quantity will be taken. The delivery date however, is not firm. Schedules could be pulled in or pushed out (within contractual limits). Beyond the six month horizon, planning data is for the supplier's own planning purposes and is not a firm commitment. Deliveries to the factory are controlled by a "pull" arrangement based on a kanban system. Certain arrangements are made within a two month window to make adjustments to kanban arrangements if necessary. The supplier arrangements also call for a level of flexibility from the supplier to meet percentage fluctuations in schedule as shown at the bottom of Figure 5.3.

POs in the system can have a status of planned or firm. Setting orders to firm status can be used to make the MRP system recognize the periods within which quantities

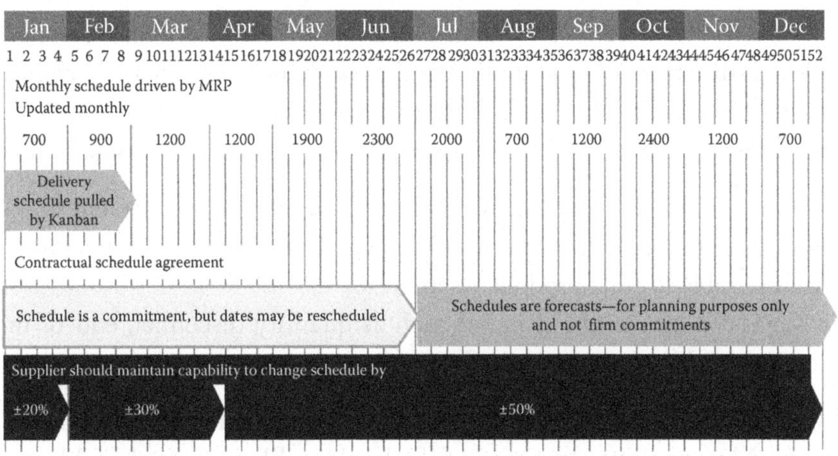

Figure 5.3 MRP schedules, pull arrangements, and contractual terms.

should not be changed. However, take time to understand how your particular system behaves when, for example, there is additional demand before the due date of an existing firm order.

5.3.1.1 MRP Results There are a number of ways sensible, good management checks can be arranged to pick up planning errors:

- Scheduler/commodity manager briefings should be held after the S&OP approval meeting and before MRP output is ready to be actioned. In these meetings, the S&OP manager should summarize changes in levels of demand or engineering specification.
- A small number of items can be selected as *schedule tests*. We should be able to check if their demand schedules reflect the new S&OP plan. They are unlikely to be exactly the same as the top-level S&OP plan because lead-time offsets, batching, and inventory adjustments will be brought into the equation. However, it is possible to select some that give a useful check if things are right. If the *schedule test parts* are toward the bottom of the BOM, they will provide some checking of the BOM and planning higher up (although they will be more complex because of higher-level batching effects).
- Value checks can be used to indicate if there are significant errors in supplier schedules. Many systems provide a *commitment report* facility which values the schedules that are placed or about to be placed. Tracking these on a regular basis can indicate unexpected changes that may lead to discovery of planning or PO placing errors.
- Suppliers can help to guard against effects such as the Forrester effect (see Section 5.3.2).

5.3.2 Forrester Effect

Forrester defined this effect in the 1950s and a lot of useful work has been researched and published since then. When applied to the way MRP systems work, the effect can introduce large variations in demand as planning descends through the BOM.

Large changes in demand can be introduced because of a wide range of factors:

- Planning parameters that control the way order batch sizes are determined
- Lead times at successive levels of supply (e.g., from raw material to component to subassembly to final product assembly)
- Artificial changes in the demand such as quantity discounts, end-of-month discounts, and buy one, get one free promotions

Jay Wright Forrester worked on the dynamics of a wide range of systems as a professor at the Massachusetts Institute of Technology Sloan School of Management. He was also a pioneer in many areas of computing that we now take for granted.

The destabilizing effect in the supply chain known as the Forrester effect is widely recognized.

The effect of these issues working together in the planning system is to distort the quantities that suppliers are asked to provide. Typically, the effect causes very high requirements followed by very low (or zero) requirements. Keeping planning systems simple will help to stop the effect, but a good safeguard is to provide the supplier with the top-level schedule as well as his or her own schedule of demand. The supplier can then recognize the effect.

5.3.3 Instability in MRP

Many talk about MRP becoming *unstable*. This means that successive outputs are so significantly different that they become unmanageable; for example, new operations plans or supplier schedules cannot be met because of capacity and time-scale constraints. It is important that changes to S&OP plans are made carefully and that plans are *achievable*. S&OP managers and approval teams usually avoid making too-large changes to top-level plans, particularly in the short term; otherwise, plans will not be achievable. In order to do this, many businesses add special capacity planning bills (often added as part of the normal product BOM), which contain estimates of key capacities. These can be internal and external. MRP results can be read directly for these as a forecast of the required capacities. They can even be time-phased using appropriate lead times.

Instability can also be introduced by changes at a lower level in the BOM in combination with planning parameter effects. One of our clients made large metal components that were supplied to a large aerospace business for machining and then assembly to a key aircraft assembly. The higher-level assembly schedules are very stable, but our client experienced large short-term schedule changes. Investigation revealed that the changes were being caused by the way their customer's MRP system reacted to changes in demand and supply. For example, three components were scrapped in a batch that was being machined. The MRP run recognized the shortfall and raised an order for immediate delivery—which was unnecessary, but furthermore, raised an order for 30 components because of a batching rule. MRP then descheduled the next order because the new order for 30 meant it was now not needed until later. MRP was becoming unstable because the new demand for 30 immediately was unachievable and yet later orders were descheduled on the assumption that it would be delivered.

Work orders (WOs) of some form are what drive most MRP system demands. Depending on how the S&OP and MRP parts of the package work, demand is balanced by the supply in the form of WOs and purchase orders (POs). WOs and POs can be set with various status flags. MRP will generally move, adjust, or delete orders if they are not firmed up in any way. Setting an order status to "firm" will usually prevent MRP from adjusting or moving it, which has the effect of stabilizing schedules and preventing supply being assumed to be available within lead time. However, check carefully how your system deals with additional demand within a period of firm orders. This can be, for example, to produce error messages or for example to raise new supply orders beyond the current FIRM orders.

5.3.4 Spares or Other Nonplanned Demand

Spares type demand is often planned using a special BOM. These are called spares bills, unplanned use bills, indirect bills, and various other names. The concept is easy to understand: include a demand for the spares bill in every month and set up spares usage items as constituents of the BOM with a suitable quantity depending on the spares usage. However, it is the management of the BOMs and spares usage that requires close attention. If the spares quantity is set too low, parts will be continually short and pulled forward. If we try to *play safe* and set the spares demand artificially high, we run the risk of high (eventually even obsolete) inventory and continual push-outs of orders from each planning run. Many people use safety stock to cater for this type of demand, and if quantities are low compared with normal usage, and/or supply is readily available with short lead time, this will work. If quantities are quite large in comparison with normal usage and the lead time is longer, a more sophisticated solution such as a combination of a smaller safety stock and spares bills can be successful. It is important to establish an active review and management process that compares the unplanned parts used for each period and adjusts the spares bills to suit.

5.3.5 Optional Features

Product variants are often dealt with by including a quantity of less than one per product on the BOM. In this way, for example, two electrical motors can be included on the BOM, both with less than one per product. Many people *overplan* these parts by increasing a typical use of a part of 80% by setting the BOM quantity per at, for example, 0.85. This has the advantage that a slight excess of the optional parts should be planned. However, if this is overdone, the issue of continual descheduling in each planning run due to underusage as mentioned in the previous section is likely. Sometimes there are a number of optional parts linked together; for example, a voltage option where several motors, starters, and bulbs are dependent on the voltage package required. In these cases an optional module in the BOM is often used quite successfully. This is illustrated in Figure 5.4.

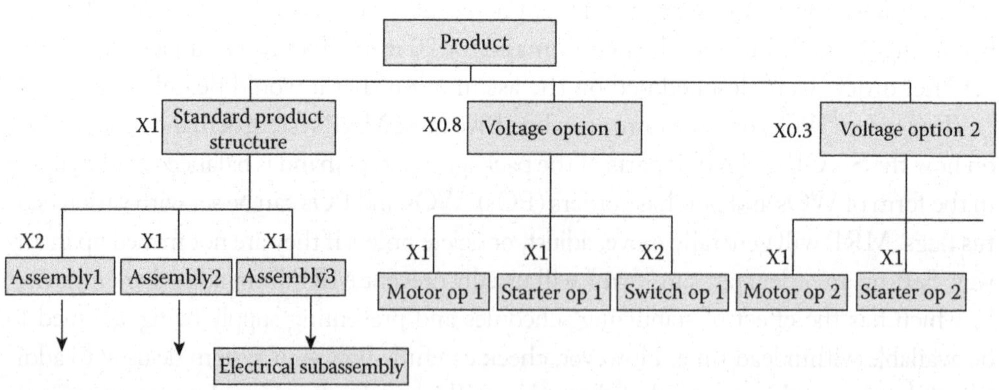

Figure 5.4 Optional features on a BOM.

5.3.5.1 Phantoms, Purchase Flags, and Shop Stock Flags A phantom flag set against a part indicates that although it is a logical part of the construction of the product, it is not a part that is planned to exist and be physically recorded by the system. A frequent use for the phantom flag is to define parts being assembled together from the engineering perspective, particularly where they need to be drawn for some technical reason. The phantom stage will be made as part of a larger assembly and not recorded as an inventory item. The planning and inventory systems will ignore it. None will be planned and parts called up on pick issue sheets will include the phantom's constituent parts. However, if there is stock of the phantom for some reason (perhaps the engineering process required a certain number to be made that were not used), the inventory will be taken into account by the MRP planning run, and also the pick issue sheet system. This leads to a trick sometimes used to ensure that all existing inventory of a component is used before the new part is introduced. To achieve this, the old part is made a phantom and the new part is made a constituent part.

A *purchase flag* indicates that a part is purchased out complete and that the constituent parts are provided by the supplier. MRP planning runs will therefore ignore all constituent parts. Many systems also have a subcontract flag that allows all or some of the constituent parts to be supplier supplied, but also, if required, issue some parts to the supplier.

Shop stock flags vary slightly depending on the system package, but generally allow for parts such as fasteners to be supplied to point of use by a supplier without the need for any purchase schedule. Planning for such items is still carried out because it provides a good check on the quantity that the supplier invoices us for over a period. No POs are generally recommended if this flag is set, but planning results for these items are usually routed to a special buyer group.

Probably the most important *guru* in the development of MRP was the late Oliver Wight, whose books are still a very valid and instructive read. One of the important

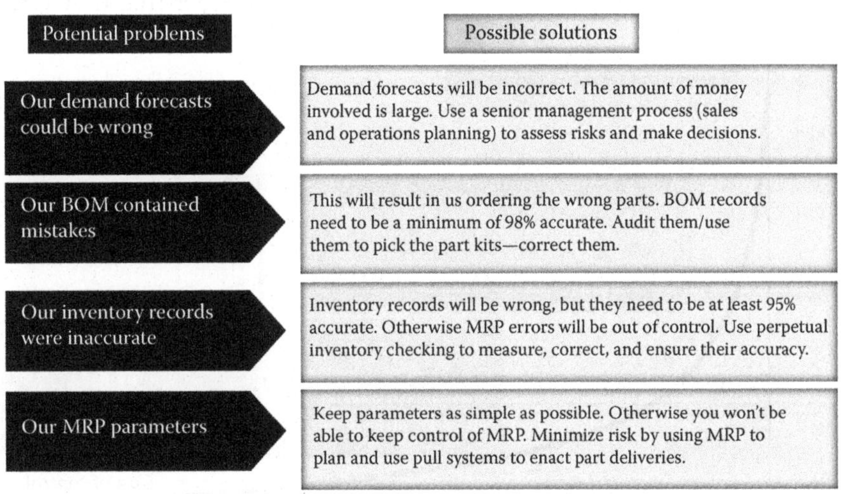

Adapted from Oli Wights's "A Class" MRP criteria

Figure 5.5 Problems with MRP.

things he realized was that while MRP is one of the most useful and powerful computer-based facilities, it is also dangerous. It can very quickly cause as big a mess in the supply chain and operations as almost anything. Oliver Wight's list of criteria for *A class users* includes guidelines for the some of the main and most dangerous causes of failure. Figure 5.5 shows some potential problems and solutions.

MRP is an extremely powerful tool. We would find it difficult to manage supplies to most factories without it. MRP can also cause the largest mess more quickly than any other system I've seen. Keep parameters simple and experiment to fully understand how it behaves under various circumstances.

* * *

5.4 Making Logistics Easier

Within logistics, the planning of parts supply is complicated, always going wrong, and can cause expensive errors. Sometimes, one of the key issues is the sheer amount of work to do. There are so many parts. Their situation is constantly changing due to fluctuating customer requirements, problems with supply, engineering changes, and so forth. Simplification is usually highly beneficial. Here are some of the ways to do that.

5.4.1 ABC Parts Management

Almost everyone uses some sort of ABC approach. The point is that even though the cost of the components at the C end of the curve is getting very low, the cost of controlling each part can be almost as high as an A class part. We therefore try to reduce the number of POs, invoices, and receipt/issue transactions for the lower-cost parts. As Figure 5.6 shows, A class parts are so few in number that they can almost

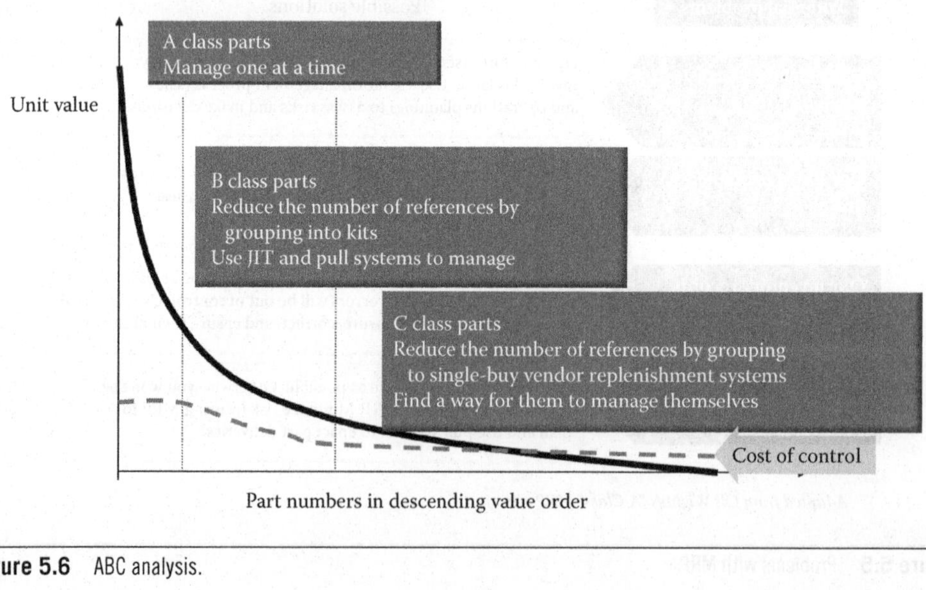

Figure 5.6 ABC analysis.

be managed on a *personal* or one-at-a-time basis. In one position I had a report printed twice a week showing the status of A class parts—inventory, POs, allocations, and engineering change notes (ECNs).

B class parts are not insignificant in cost, but still too many to manage closely. We like to reduce the management time taken while still remaining in control by two particular techniques: kits and pull systems. These are discussed in Section 5.4.2. Although the cost of control is drastically reduced, the accuracy and cost of control achieved is not.

C class parts cost very little, usually cents. However, the shortage of only one C class part can stop an operation. Keeping too many of them is not costly and so we do not mind if the control is slightly less tight—as long as we do not run out! Several supplier solutions to this are discussed below.

5.4.2 Kits

Kits can reduce the number of parts to be managed by approximately six to eight times. This applies to field operations, service, manufacturing, and even to retail supply chains. Jobs carried out in the field, for example, in an oilfield or at a construction site, often use the same sort of parts repeatedly but parts are planned and controlled individually. The specifications for carrying out the work sometimes even list the parts to be used and require them to be checked by the engineer before going to site. In service work, again the parts used on a job are often the same. We hear all sorts of reasons why it cannot be done—some of the parts are only used sometimes, some of the parts are for *service exchange*—but we see many elegant solutions to using kits in the service industry. Figure 5.7 shows some key aspects of a kit ready for maintenance that might be needed at a field operations job. An excellent solution used by one client is called *ready boxes*. The boxes themselves are made to suit the parts for different service kits.

This is not as expensive as it sounds because the boxes are standard sizes with customized foam trays to hold the parts and materials. The ready boxes are held in

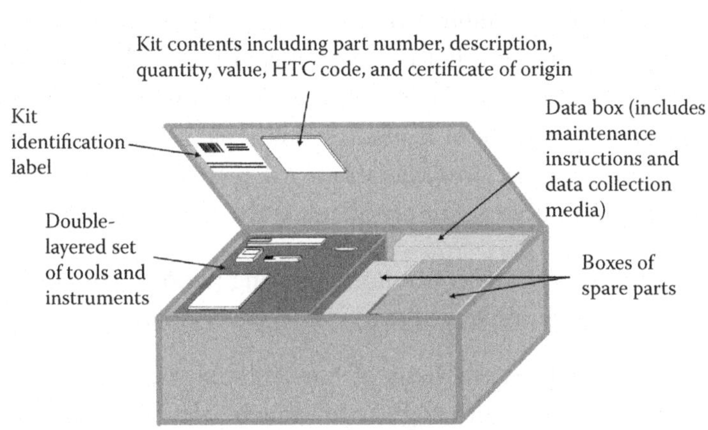

Figure 5.7 Kit box ready for an operational job.

stores to be picked up by the engineers who use some of the parts and then return them to the box. The boxes are then returned to stores, but here's a clever supply chain bit—a *ready box supplier* collects the used boxes and returns them to stores ready for use again. The *ready box suppliers* make some of the parts themselves, but most are obtained from other suppliers. An example we have seen that goes even further is in aircraft repair. A client uses kits that work much the same as the ready boxes but also contain a CD-ROM containing instructions for diagnosing and correcting the fault and for collecting data about the reported fault and its repair. In manufacturing, the supply of part kits is relatively well established. One with a slight difference that was very useful to a client is the supply of all small specialist parts for engine assembly stages in special trays that sit at the assembly station. Again, the supplier of the trays to the assembly station obtains most of the parts from other suppliers; the point is that the controllable parts have been reduced for our client by about 10-fold.

5.4.3 Pull Systems

Pull or *kanban* systems come in a wide variety of shapes and sizes. Some form of signal informs the supplier that some of his or her product have been used, which triggers the supplier to replenish them. The signals can be

- *A ticket or tab removed from the part before it is used and located somewhere for the supplier to find.* The supplier replenishes the used quantity with tickets attached so the process can continue.
- *Boxes of parts.* When the parts are used, the empty box is collected by the supplier, refilled, and brought back again.

It is important that after the supplier has collected the kanban signals (tickets, empty boxes, etc.) the number of parts left available for use is enough to last until they are replenished. If the supplier visits twice each week, the quantity left should therefore be one-half a week's demand, plus a safety quantity. To make this work, a *kanban population* (KP) is generally established. This is the total number of signals (tickets, empty boxes, etc.) in the system. For twice-weekly visits, the KP should therefore be 1 week's demand plus safety—which is one-half a week on our premises for us to use plus safety, plus one-half a week of kanbans being replenished by the supplier. There are formulas and consulting services available to calculate the KP. We do not really see that it has to be this complicated; simply calculate the KP as follows:

$$\text{KP} = 2 \times \text{expected consumption during period until next supplier replenishment visit} + \text{safety quantity}$$

Decide the safety stock on the basis of how volatile the demand is likely to be. Implement the initial system with slightly too much safety allowance and gradually reduce it. When it starts to get too low, increase it a bit! The real issue is communication, understanding, and discipline, both by the supplier and our own users of the parts. For example, the kanban tickets/boxes/other types of kanban signal must be kept carefully and located in the agreed-to place (a rack or container of some sort) for the supplier to collect. Parts should only be used for the purpose for which the system was designed—so if a new use (e.g., another product or spares demand) materializes, we need to adjust the system to include the new demands. Figure 5.8 shows an example of a kanban operation that uses the component boxes to manage stock replenishment.

Kanban systems are much easier to use for regular usage parts. The more irregular the stock items are (see Section 3.4, "Regularity of Demand"), the more safety stock needs to be held in the KP. At one time kanbans were only considered suitable for high-volume, automotive assembly type demand. However, many successful applications exist now for regular, low-volume demands. It is also not absolutely necessary for the supplier to visit our premises to collect the kanban signals and replace inventory. Our message is, be inventive and practical. Use a system that works best for your application. Two examples of remote applications are

- *Faxbans.* For example, a box of parts is emptied and needs replenishment. The box contains a sheet of paper with the part and kanban details. This is faxed to the supplier and the box and kanban sheet are placed together ready for collection.
- *Webcams.* A small quantity of a large component are located on the floor next to an aircraft assembly area. The components are always located in the same place and the floor is painted with the shapes of the required number of kanban components. The supplier, who is on the other side of the Atlantic to the customer, sees when there is a new empty space and dispatches a replacement.

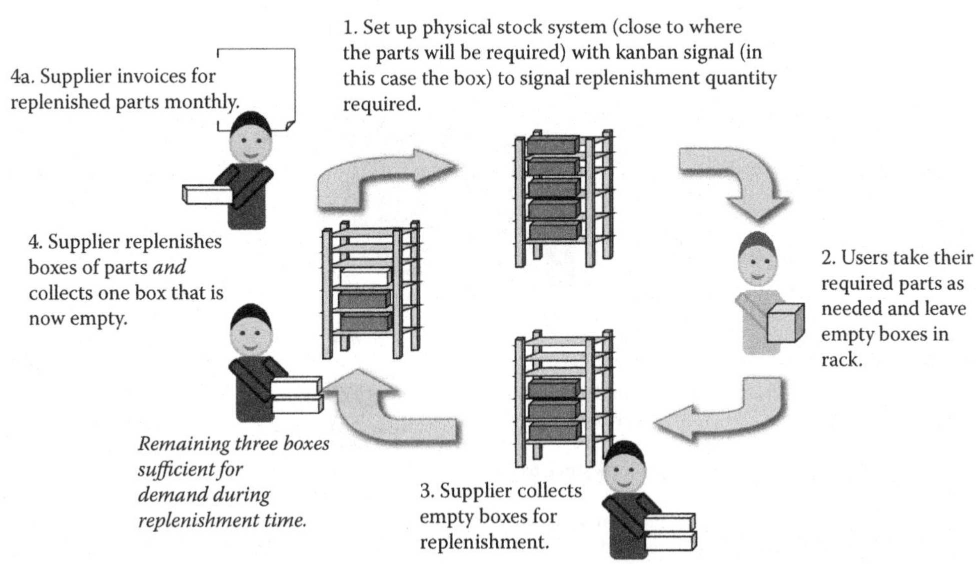

Figure 5.8 A simple Kanban loop.

5.4.4 Bin Stock Replenishment

Bin stock replenishment is the common way to manage low-cost items that are in reasonably regular use. Figure 5.9 shows a system implemented in an assembly area in India by my colleague Crispin Brown.

This is a form of VMI. Small plastic bins mounted in some kind of rack are usually positioned close to the *point of use*. There are likely to be several racks of bins set up around a location. The supplier replenishes the stock in the bins each time he or she visits—for example, twice every week. There are no inventory recording transactions for these items and the supplier typically invoices at month end for all replenishments made. A satisfactory check on the accuracy of the quantities charged for on the monthly invoice can be made from the MRP run (see Section 5.3.1, "Planning, Scheduling, and Pulling"). This system is simple but, as with nearly all kanban systems, it does rely on understanding and discipline from everyone with access to the bins. This means almost everyone in operations and a good few in other departments. Where this discipline does not already exist, it usually takes a while to get it established. It usually goes a bit like this:

- You make the arrangements with the supplier, sell the idea to finance, explain and demonstrate the system to the operators, set up the physical bins, and make arrangements to start.
- At the end of the first week, you come back to find that everything is a mess. Some of the bins have no inventory, some contain a mixture of different parts, and some of the bins are no longer there. You straighten everything out and

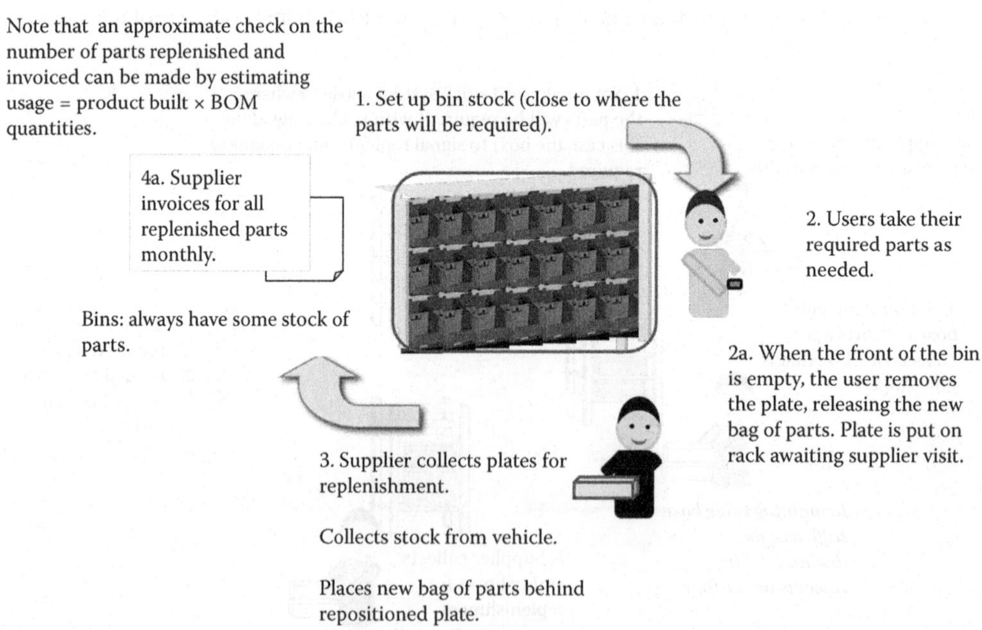

Figure 5.9 Bin stock replenishment.

you give everyone a pep talk about how they need to take the system seriously or it will continue to get in a mess.

- At the end of the next week when you come back, it is not quite such a mess, but you clean up again.
- You come back at the end of the month, and the system is in a mess again so you sort it out and tell everyone we are through with clearing up the mess and are now depending on them to keep it straight. You get a couple of team members to take ownership of checking the system and the clearing up if needed. You also find that the invoice charges are far higher than the planning and output figures say they should be. You also tell the team about this and make it clear that proper checks and balances have been set up as part of the system, and we know when too many of the low cost items have been used.
- The checks, mess, and clearing up carries on for about 6 months but becomes less of a problem each time. The team members help to keep the system working well and in the end, peer pressure is one of the things that persuades the last undisciplined system-breakers to mend their ways.
- You continue to check and support the system, which is now working well.

This sort of process is common to many changes to the logistics (and most other) processes. It is important that those responsible for implementing it do not walk away too early. Expect it to go wrong and be ready to correct it as a normal part of ongoing implementation. Also make sure that when a system like this is put in, you also implement proper checks on its correct operation and use.

5.4.5 Vending Machines

Vending machines are becoming more common, particularly for items of moderate value and with medium-volume demand. We see them in airports where you can use your credit card to buy electronics of quite high value. Many hospitals use them on wards where the nurse in charge can use a personal identification card to obtain items where some level of security is needed. In industrial use, we have seen them working very successfully for tools in a machine shop and for office equipment consumables such as toner cartridges. A shown in Figure 5.10, the machine manages who can obtain goods, calls for replenishment of items needed, and enables invoicing at the end of each month—complete with a breakdown of who had what for inquiry purposes. The machines are moderately expensive and if the supplier is going to fund it he or she may need to charge slightly higher prices. If we (the customer) are to fund the machines, we too will need to justify the cost. It is certainly possible to achieve savings through a combination of

- Reduced stores transactions and space.
- Reduction in *motion waste* because the vending machine can be located closer to the point of use.

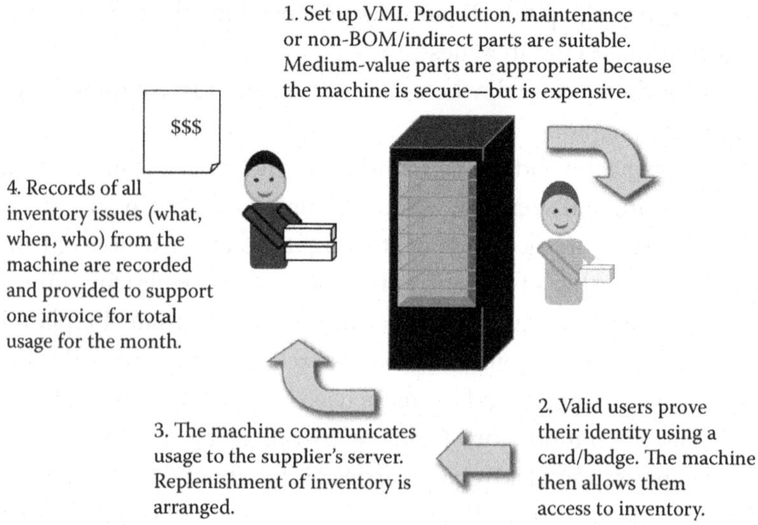

Figure 5.10 Vending machine process.

- Reduced inventory (this can be from within stores, but also from many other areas where a supply is kept just in case). This unofficial inventory is a large potential saving, but quite difficult to quantify—try a spot survey!

A favored way to finance the machine is through a reassignable lease. The lease is transferred to the supplier, but if the service is to be re-sourced, the machine lease can be reassigned to the new supplier.

5.4.6 VMI

VMI has already been mentioned in one form in Section 5.4.4. A variety of these methods are also very successful in higher-value and more sophisticated form. A client supplies a major food manufacturer with packaging for a high-volume line. The customer provides a web-based facility to enable the supplier to manage the inventory that is on a customer's site awaiting consumption. Figure 5.11 shows the type of facility provided. The customer maintains a daily schedule and the current inventory on-site. The supplier maintains the line of planned deliveries for each day. The system calculates the projected inventory and compares this to the target level of between one

		T	W	T	F	S	S	M	T	W	T	F	S	S...
Production plan		1200	1400	1450	1400	1400	800	1500	1500	1550	1550	1500	1400	1200
Deliveries		1200	1200	1400	1500	1200	1200	1200	1200	1200	1200	1200	1200	1200
Projected inventory	1150	1150	1000	950	1050	850	1250	950	650	300	50	−350	−550	−550

Figure 5.11 VMI projected inventory balance.

Where is it?	Who owns it?	Who manages it?	
Customer	Customer	Customer	Inventory is delivered, we have it, we pay for it
Supplier	Customer	Customer	Customer furnished equipment/inventory
Customer	Supplier	Customer	Consignment inventory
Customer	Supplier	Supplier	VMI Vendor-owned, VMI
Customer	Customer	Supplier	VMI Customer owned, VMI

Figure 5.12 Management and ownership of inventory.

and one and a half times the next day's demand. In this case, the customer appears to have increased his schedule and the reduction in projected inventory falls below target on day eight. The supplier's planner needs to address the issue within 1 day.

VMI inventory is usually owned by the supplier until it is used. *Consignment stock* is to some extent the predecessor to VMI. *Consignment stock* is also owned by the supplier until used, but the responsibility for replenishment management sits with the customer. The customer needs to tell the supplier when to replace inventory. It is of course a cash flow advantage if the supplier retains ownership of the inventory while on our premises until we use it. However, we do see many large businesses with good cash availability asking much smaller suppliers to do this. We think this logic is questionable. Figure 5.12 shows some of the options possible for inventory ownership.

The logistics mechanisms described in this section have two major benefits:

1. *The administration effort and cost is much lower and very appropriate for lower value items*
2. *The accuracy of deliveries is a more accurate match to demand*

Implementation requires patience and persistence because it relies on the discipline of its customers.

* * *

5.5 Inventory Management

The storage, custody, and distribution of inventory as a function is undervalued in most organizations. The value of goods passing through these functions is usually well over half the cost of goods sold (COGS). Inaccuracy and damage can lead to shortages, and slow or inaccurate distribution can damage customer service. Businesses where this level of impact is possible should have good strategies, proactive developments, and measures for their logistics functions.

5.5.1 Inventory Reduction and Optimization

All inventory is bad—unless you need it! The correct level of inventory is a balance between cost and other issues such as risk and customer service level. We have worked with client teams to deliver large inventory reductions on many occasions. We typically reduce inventory levels by 25% while increasing customer service levels. It is important to couple inventory reduction to customer service improvement; otherwise, it will be assumed one is at the expense of the other. We use the following three-step approach:

1. *Question and (if required) change the customer service level.* For example, we often do an analysis of the value of inventory held for a customer and compare it with the annual profit made from the customer. Where the inventory value exceeds the profit, action is needed! Having decided an appropriate level of customer service, if we fix the planning approach and parameters, the service level follows.
2. *Changing the planning approach.* This is summarized in Figure 3.11—Section 3.7 Forecasting and Planning for Different Types of Demand. Depending on the demand and part characteristics (volume, value, regularity, dependency, criticality) determine what is the best control approach (MRP, VMI, Kanban, etc.).

   We are usually able to get the data we need for this stage from the planning and inventory systems: ABC analysis shows us the relative value and also volume, runner/repeater/stranger analysis shows us repeatability, MRP can tell us the dependency, and we can split new product and spares demand. We usually present the analysis to a senior team and agree to classify products and parts according to this analysis.
3. *Optimizing the parameters with the planning approach.* When we have agreed on the classification of parts and the appropriate planning approach, we need to optimize the control parameters. For example:
   - For kanban-controlled items we need to optimize the timing of replenishment visits and the *Kanban Population* (KP). We can apply formulas to do this and although this is useful, we usually start by splitting parts by value:
   - For A class items we look quite carefully at demand variability and the lead time. If we can replenish reliably on a daily basis, we can set the KP at 2 days' consumption plus some safety for variability in consumption. This approach will work as long as we can replenish within a maximum of 1 or 2 days, which may mean the supplier holding some inventory for us. For longer lead times (and consequently higher KPs), the approach will continue to work as long as usage is not too variable.
   - For lower-value parts (B class) we can afford to have longer lead times and larger KPs. We might have a 2-week lead time and a KP that allows for a 20% variability in usage, for example. The initial levels set are usually intentionally to be too safe. Over a period of time we monitor delivery

lead-time reliability, demand variability, and customer service failures by part and adjust them.

- For C class items, we would not expect to use kanban control. A two-bin system would be more appropriate.
- For parts controlled by MRP schedules, we need to optimize the control parameters. A good example of this would be the use of a parameter where MRP will raise an order to cover requirements in a specified forward period (e.g., 1 week, 2 weeks, 4 weeks). This parameter is controlling the frequency with which the supplier is asked to deliver, plus the level of safety stock we will aim to hold. The safety stock is set depending on the variability of demand and reliability of supplier deliveries. This is also set (intentionally) too high, and gradually reduced as we measure the results. The period of order cover is really dependent on the cost of order raising and delivery. The lower the value, the more we should order at one time (longer order coverage). An economic batch quantity calculation can be used here, but it is often quite obvious what can be achieved from discussions with supply chain people and supplier.

We work through this process with the planners and other supply chain team members to ensure they understand and agree with the approach. It is important to have good data and use it as the logical basis for deciding the planning approach and setting the detailed planning parameter levels. This sort of approach can be carried out with or without external help. It is important to have a structured approach and to have good data so that the work will continue after the initial exercise has been completed.

Optimizing (or even deciding not to hold some inventories) can deliver very large cash benefits and often has a side effect of improving the efficiency of the planning, shipping, manufacturing, or supply activities involved.

* * *

5.5.2 Warehouse Layout

Warehouse layout is a Lean tool that is focused on the reduction of journey waste within the warehouse. In principle, high-level pick parts (and heavier) stock should be kept closest to receiving and dispatch.

Figure 5.13 shows a simplistic store layout with receiving and dispatch at the same end of the building and locations arranged with higher numbers being farthest from them.

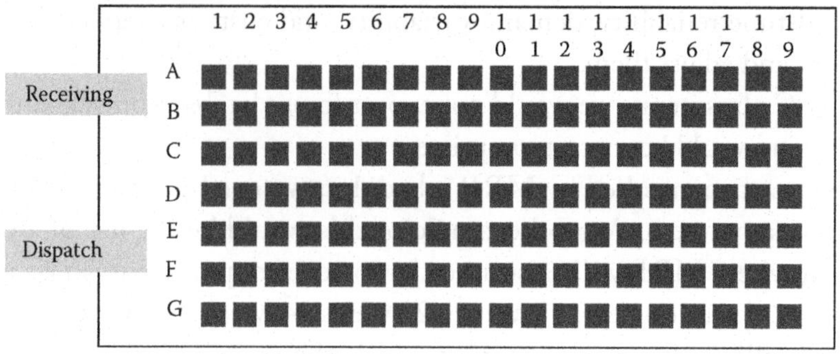

Figure 5.13 Warehouse layout.

Draw a map of the physical store's layout and plot the journey routes for putting away and picking stock (and any other significant journeys). Establish the relationship between the length of journey and the location number. This needs to be a number representing the length of journey to and from each location or group of locations.

Strip part references from the inventory system along with a summary of the number of receipt and issue journeys for each part. If there are significant weight differences or some picks are for more than one pallet (and require more journeys), factor this into the receipt and issue summaries.

Add the length of journey data from the map for each location. Add a field that multiplies the journey numbers by the journey lengths to show the total journey distance, cost per month or week, and so forth. Use the spreadsheet to replan locations to relocate parts and reduce journey length and cost.

Automation in warehouses is now common. A wide range of storage racking is available, into which automatic guided vehicles (AGVs) or some form of moving crane device can store and retrieve inventory. Racking can be fixed or movable and height can be increased significantly with automation. This is often one of the major cost advantages used to justify the investment in automation along with the reduction in manpower.

Initially, automated warehouses were mostly based on handling of complete pallets. Now it is common for locations or containers to be presented by the warehouse management system (WMS) indicating how many parts should be picked from them. Goods from a number of pallets, containers or locations presented are manually or automatically transferred to new *mixed pallets* ready for dispatch to customers. Individual cartons of goods such as drugs can now be automatically stored and retrieved in, for example, medical stores.

Paternosters are storage devices often used for storing smaller items taking up relatively small shelf space. Shelves are mounted on a frame that rotates inside a cabinet, moving to present the required location to a person at the *picking face* for location or retrieval of goods. Items can be retrieved manually, one at a time, or more commonly by loading a picking list—for example for a batch of PCB assemblies—to the paternoster. The machine will then present the required shelf locations in sequence. Figure 5.14 shows the paternoster rotating in order to present the requited items for picking.

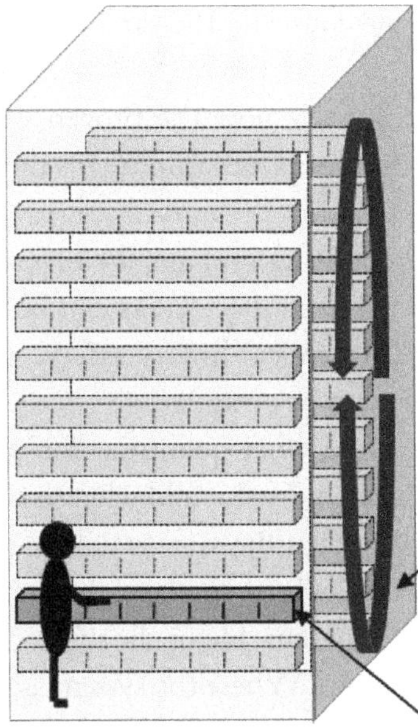

Figure 5.14 Paternoster storage unit.

The design of automated warehouses is too specialized for us to cover in this book and there are specialist companies who will design, and if required, install automated warehouses.

Warehouse management systems (WMSs) are said to be included in some ERP packages. In reality, these are usually inventory recording systems. They record what inventory part codes and quantities are located in warehouse locations. ERP system facilities interface well with BOM and part records, which is particularly important for manufacturing and other operations stores, but not so much for retail. Receipts and issues in ERP systems also interface well with POs and customer orders. Facilities for *cycle counting* are also included in most ERP packages. However, for large automated warehouses, particularly those serving retail customers, a WMS is often used and interfaced to other facilities (e.g., for receipts and issues). The advantages of the dedicated WMS are

- Interface to automatic machines
- Determine the best location for arriving goods
- Optimize retrieval routing

5.5.3 Inventory Accuracy

It has been proven that the old approach used to check the accuracy of inventory records—closing the store for a day and counting everything—can introduce more errors than it uncovers. This can happen for a variety of reasons: counters get tired, counters may not be familiar with the parts and mistake their identity, there is little time to investigate stock errors, or several people counting the same area may double-count or miss an area.

Cycle counting or perpetual inventory checking is now the proven way to obtain high levels of inventory accuracy (e.g., 98%+). As the names imply, this method counts a small area of the store each day.

I prefer to divide the store into sections for counting and get the inventory recording system to ask for certain areas to be counted each day. Areas can be allocated to storekeepers for accuracy and other issues such as cleanliness, safety, and tidiness. Storekeepers can count their own areas or they can be swapped.

When counting an area, the operator is not told what quantities for each part the system currently records. In fact, I prefer him or her not to be told what parts the system records are there. That way, parts that are missing from the system will be picked up.

The operator or counter will typically count a rack of inventory and then return to a terminal and access the cycle-counting facility on the system. He or she will sign-in and tell the system which parts and quantities have been found. Where the system agrees with the part code, location, and quantity, a successful count is recorded. If the system and count quantity don't match, the operator is asked to re-count. If however, the re-count quantity does not agree either, a supervisor check is called for. If the supervisor check confirms the operator count, the system quantity is corrected and an inaccurate inventory record is logged. After this, there should be an investigation of the cause of the error. A good store management will then try to change procedures to prevent the cause of the error from recurring. This way, both the inaccurate record and the cause of the problem are addressed.

Many people set their system up to support more frequent counts of A class items. This seems sensible and can be made to work by grouping valuable parts by location.

At one time, accuracies of around 80% were commonplace, but now, accuracies in the high 90s are necessary. Done the right way, the improvement of inventory accuracy by store personnel themselves is motivating. The levels of accuracy that can be achieved by motivating people and using cycle counting are quite astounding.

The impact of good inventory accuracy on customer service level and inventory value delivers significant business benefits.

* * *

5.6 Distribution

In many businesses, distribution is becoming a major cost and a major customer service issue. Manufacturing businesses have worked on their production costs for years, they know the cost of each operation in detail, and they measure the performance with exacting key performance indicators (KPIs). These days, distribution is often an important area for cost reduction. Given a little work, prices for most forms of transportation can be very competitive. However, to optimize cost and customer service level there are more detailed questions that need to be answered:

- What delivery services do customers want (note that this can change quickly and it may be worth doing a go to market [GTM] survey) (see Section 2.1 Understanding Customers and Demand)?
- Where are the best locations to distribute from?
- What inventory should be held in each location?
- What are the optimum routes for distribution?
- What are the optimum modes of transportation to use?
- How can we optimize tactical routings each day?
- Which parts of distribution should we outsource and which should we do ourselves?
- What is the cost of different levels of customer service and how should we optimize this?

Financial models are usually essential to get to the right answer and these need to be constructed to meet the needs of the specific question. In most cases, this includes the optimization of customer service level, distribution costs, and cost of holding inventory.

Spreadsheet models can be used if the situation is not too complex. For example, a business manufactures products in one factory and sells them through distributors in the same country. If there are not too many of them, the distributor locations can all be set up on the spreadsheet together with the number of loads delivered. Distances between alternative distribution warehouses can be entered and total truck distances traveled can be calculated. The costs and associated times for the alternative locations can then be calculated and compared. This will enable a simple comparison of costs and an assessment of the customer service level that could be achieved.

Costs of warehousing (buildings and people costs) vary significantly from one area to another and these costs also need to be factored into the model.

Quotations from alternative transport providers can also be compared by spreadsheet. If the RFQs are well structured, alternative warehouse locations can be compared at the same time as transportation providers. In effect, the specialist logistics companies will do the distribution analysis. You have to make sure the supplier quotations give sufficient information to enable you to select the right solution. For example, your quotation could ask for a breakdown of costs by region and for different modes of transportation and service level, and if distribution warehouses are to be used, a breakdown of their costs (buildings, people, etc.).

A wide range of specialist distribution software is available and varies in complexity, capability, and of course cost.

A simple example of the use of specialist software would be similar to the spreadsheet above, but with the potential for more locations and a distribution network. The mileages between potential positions of warehouses and the customer base would be known by the system. This enables options to be evaluated more easily—for example, system facilities would enable you to move main and distribution warehouses to new locations. The system would then recalculate mileages, cost, and customer delivery times for each set of locations. Software packages will also enable you to experiment with different load densities in different vehicle types.

In addition to the analysis supporting the strategic solution, you can also use software packages to optimize tactical (daily) distribution. We know a business that distributes daily on a national basis from 12 factories—each producing a range of fresh products. When a production problem occurs, limiting the output of one of the factories, other factories make up the capacity shortfall. Their software package can then help them work out the best way to trunk the product in new quantities from each factory. In these cases that require more accurate routing optimization, journeys will often be calibrated for the system by recording mileages, routes, and times from on-vehicle telemetry. This can also enable the software package to take into account peak traffic times and delays.

Managing distribution is demanding because it is a complex and detailed task, and one that changes quickly. For example, the distribution work for a particular day may have been well planned to utilize 46 of our fleet of 50 road vehicles. Two of the vehicles will be off the road for service today. Early in the day, two of our drivers call in reporting they are sick. We call a number of standby drivers and arrange for two of them to cover. One of the manufacturing centers calls to inform you that a load to be delivered to one of the most distant locations will be available 2 hours later than planned. You change the schedule, but one of the drivers cannot manage the later return to depot and you have to switch him with one of the standby drivers. You then contact the customer service representative to contact the customer and explain their delivery will be late. One of the vehicle tractor units is on route with its load, but has developed a fault. The driver's description of the fault suggests the vehicle needs to be recovered and you make arrangements to do this. You now need to find another vehicle but the maintenance garage tell you the two being serviced cannot be made available in time to help. You manage to hire a tractor unit and send another driver to collect it. He will then drive to where the vehicle has broken down and take the trailer unit to make its deliveries. You need to change drivers around so that the one returning from the broken-down tractor unit takes the load the replacement driver was to have made. This is a fairly typical pattern of events for the first couple of hours of the distribution day, and it may get better or get worse. This challenging task requires very active engagement of managers with a flexible and capable workforce.

More complex software can be used to optimize multimode transportation that can be cross-border and include stages of sea, air, or road freight, importation, customs clearance, and even some taxation issues. These packages can be used by businesses who, for example, manufacture in China, then ship to Europe and the United States for further stages of distribution. Figure 5.15 shows an example of the graphical form of output for a similar example of distribution into Europe.

In this example, sea freight is used to ship products into Europe and into a central warehouse. Some products are distributed to customers directly from the central warehouse and some products are sent by trunking vehicles (larger, long distance trucks)

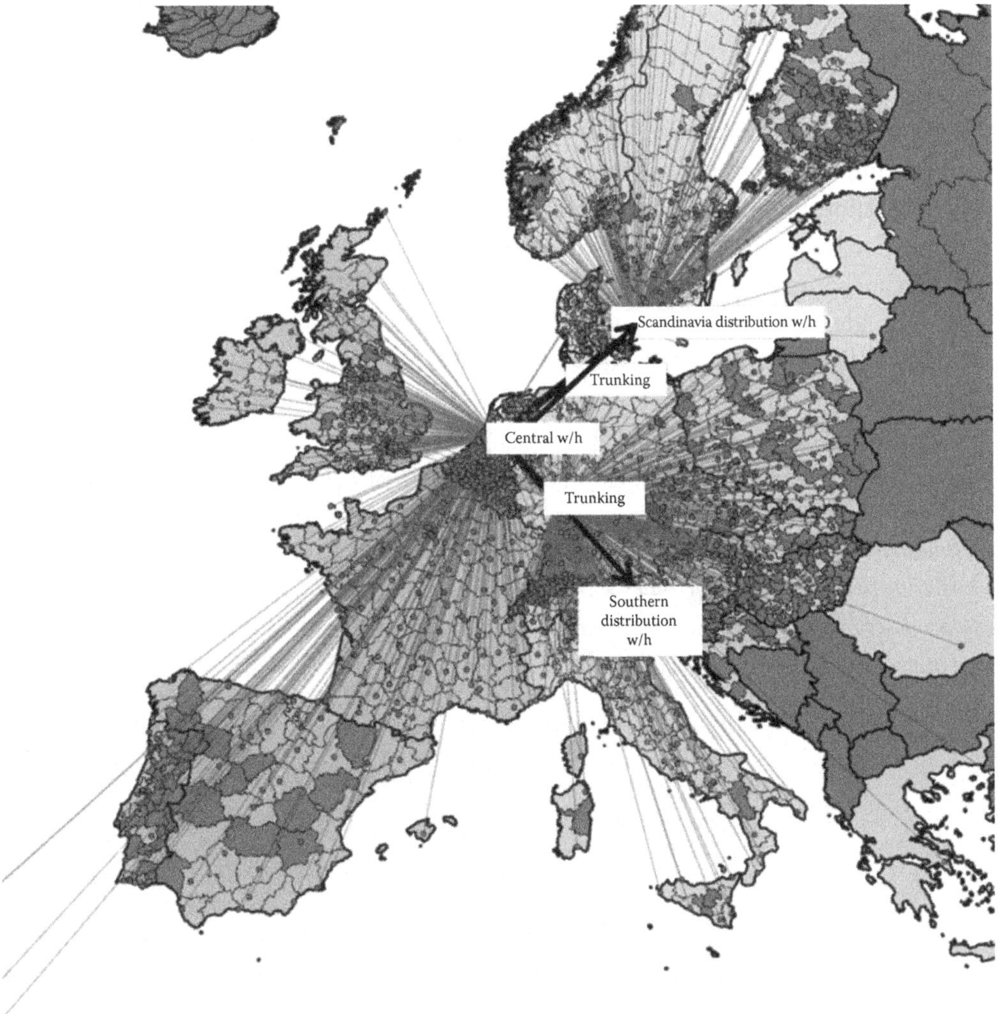

Figure 5.15 Example of a European distribution network.

to Scandinavian and Southern European distribution warehouses. In this example, a very sophisticated package was used to optimize the following:

- Incoming freight costs into a choice of ports
- Importation and local value-adding taxes
- Location of the central warehouse (local costs and distance from customer locations)
- Location of two distribution warehouses (local costs and distance from customer locations)

In this case, a large number of customer locations were grouped into clusters by the software to enable more effective analysis. This is an example of a large and complex network analysis.

Some businesses have highly variable delivery requirements that need to be optimized on a daily basis. Software is available to optimize vehicle load planning and routing. Delivery times can be estimated and preallocated time slots planned. These are frequently complemented by in-vehicle monitoring, reporting, and recalculation of, for example, scheduled delivery times where delays occur.

Distribution planning is frequently integrated with manufacturing planning and control. For example, a factory making a product to customer order provides their distribution company with the customer orders that will be available to ship just over a week in advance. The distribution company plans optimized loads and journeys to make the deliveries. The plans are then transferred back to manufacturing who are able to fine tune their plans. For example, the sequence in which some of the products are completed may be adjusted to match the delivery plan. Finished goods will be arranged by transportation load and in sequence of collection in the dispatch hall.

Distribution, particularly if it involves shipments to retailers or final users, has become an important field. It can impact sales if customers are not satisfied with service levels and it can impact costs (businesses typically spend between 3% and 10% of their cost of sales). Finding and managing the optimum solution can be a demanding task.

* * *

6

Supply Chain Business Processes

6.1 End-to-End Processes

Business processes have received a great deal of attention since business process reengineering (BPR)* became popular in the late 1990s. BPR has been used for work ranging from cost reduction to reshaping businesses and providing new focus. The high-level processes shown in Figure 6.1 were used by a management team to define and evaluate their key business processes. This business manufactures large machines customized from standard designs to meet the requirements of large contracts. Interestingly, *supply chain* is only mentioned in the business support (10) arrow. Supply chain work is an important part of many other arrows:

- Representing the part key suppliers play in strategic development (1)
- Bringing potential new technology from suppliers as part of technology development (2)
- Managing the contribution from suppliers to new product and process development (6)
- Managing and costing the supplier part of winning new business (7)
- Managing supplier's development and delivery of customer projects and orders in order fulfilment (9)

In fact, it could quite rightly be argued that supply chain has a part to play in every one of the high-level business processes. Making sure the supply chain is making the right contribution to these processes is a good starting point for success. This strategic level is often missed, resulting in the supply chain focusing on tactical work. Strategic supply chain work is likely to involve attending seminars or internal meetings about, for example, business strategy, new product development, and customer prospects. It is likely to involve building useful networking relationships, visits, and meetings to prospective suppliers that are speculative, investigative, or at most, tangible fact-finding. People doing this work must have a good knowledge of company business, but must also be creative and able to *think outside the box*. They must also be able to communicate well and share and develop their ideas with others inside and outside the business. Supply chain people doing this work will be involved in all sorts of processes.

* Defined by Michael Hammer and James Champy in their book *Reengineering the Corporation*, HarperBusiness, 1993.

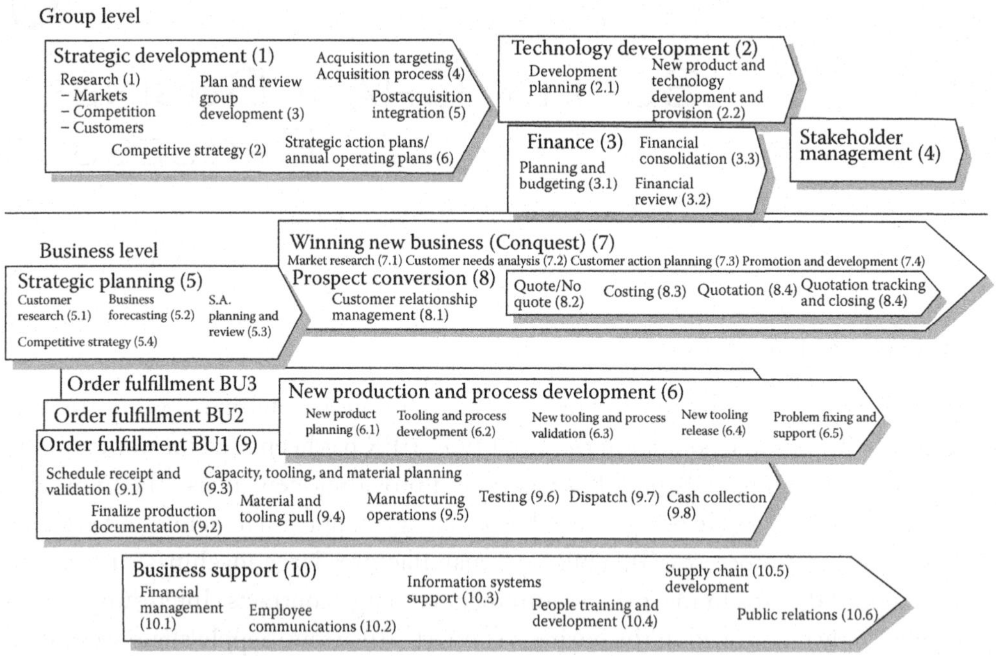

Figure 6.1 High-level view of business processes.

They will need to estimate costs, make early evaluations of potential suppliers, and arrange for potential suppliers to work with research, development, and design teams. The work is difficult to define and measure in the procedural way processes are often defined (i.e., step by step, using ratios and direct measures of success). Success needs to be measured, sometimes with an element of subjectivity regarding the contribution of ideas and the amount of progress made in developing tangible results.

It is generally recognized that end-to-end processes need to be considered, not those within a department or performing a subprocess. Terms like *innovation to cash* and *purchase to pay* (P2P or P to P) are used to represent this type of process. The transactional level supply chain processes, such as P to P, need to achieve a number of end results (e.g., having the materials we need and also having paid suppliers). This is likely to be of critical importance to the business. However, P to P can be argued to be the only part of a process that starts with forecasts and customer orders and finishes with cash and profit.

The processes shown in Figure 6.2 are from a different type of business, but the titles of the four process groups are similar. These processes are more tangible in that they can be measured practically and related to mostly transactional work. The effectiveness of these processes can be improved using *Lean tools* such as *process mapping*. There are key requirements for this type of process such as clear accountability and responsibilities, key performance indicators (KPIs), trained and adequate resources, and IT support systems. Process mapping should check all of these and make sure the process is effective.

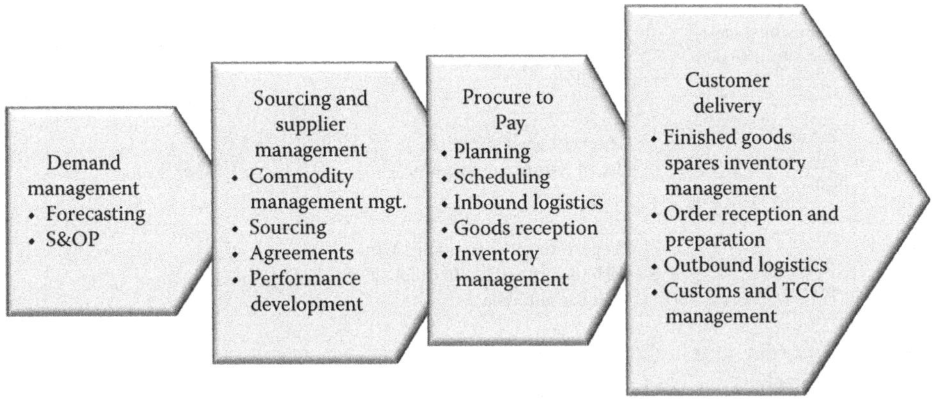

Figure 6.2 High-level description of supply chain processes.

Each time we process map, we find something surprising. These are often simple, for example:

- Orders go for an approval that is unnecessary
- A process step is done in series (holding up the process) when it should be a separate process done in parallel
- Information is supplied in the wrong form or needs to be sought from more than one source

Examination of these processes should make sure they meet their target objectives in the right time scale and are efficient. In our experience, developing procedures showing how the work should be done in too much detail is a waste of time. The paperwork becomes unwieldy and would quickly be out-of-date—which would matter if they were ever read. A better approach is to prepare a higher set of process definitions and enable working teams to adopt and refine their own detailed *ways of working*. This even applies to large multinational businesses where the processes are used to provide consistency across different offices. Sufficient uniformity of process and results (KPIs) can be achieved using very concise documents.

An example of this is shown in Figure 6.3. This represents the way suppliers are sought, prepared to be business partners, managed, improved, and ultimately, perhaps discontinued.

Each phase has a *phase gate*—a set of test criteria that must be satisfactorily completed before the program can continue to its next phase. A single sheet of paper describes the next level of detail for each phase and is sufficient to provide a consistent and rigorous process at the operating level.

Individual teams can usefully use more detailed guidance about their processes using a *workgroup notice-board*. Daily process details such as the number of orders received and processed and any associated issues or holdups are often displayed to help

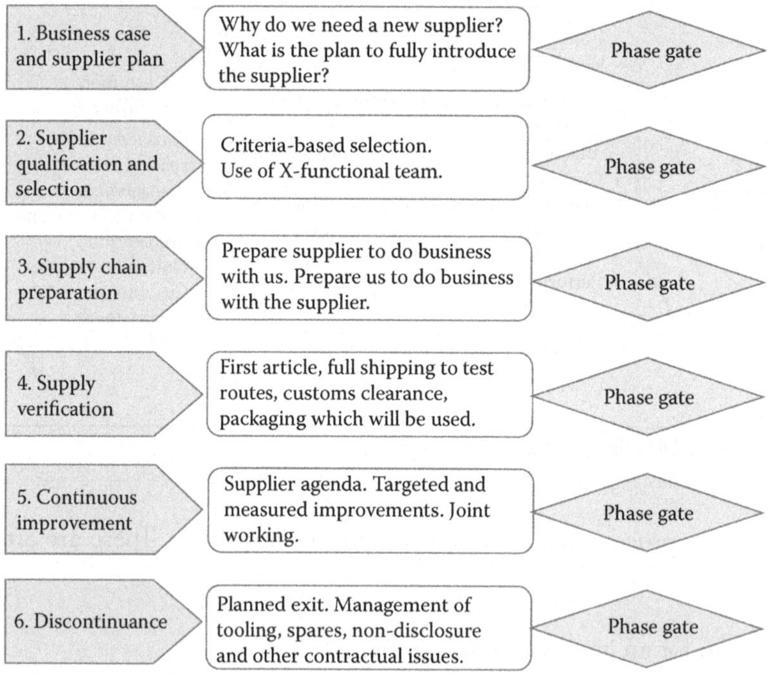

Figure 6.3 Best practice example: supplier life cycle management process.

those in the team, adjacent teams, and management to understand *at a glance* what the situation is and where an intervention may be required.

Take an end-to-end view of business processes—don't accept departments as boundaries. Do the process mapping with the people who do the work. Focus on what the process should deliver.

* * *

6.2 Supply Chain Information Technology

The introduction of new types of IT will continue to accelerate, and the ways we collect, store, analyze, and communicate information will be revolutionized again and again. So will the supply chain business processes. Selecting and implementing the best package solution is no longer enough. Those who combine a range of other devices and code to deliver the best processes will be the winners. Supply chain processes are critical to the speed and effectiveness of the business. Many have a direct effect on customer service and they should be at the forefront of developments. Supply chain people need to be among those who are looking at new technologies to see the opportunities for developing new supply chain processes. I have no doubt that this section, and in particular the IT examples, will be out of date quite quickly, but the message they deliver will still be relevant.

Supply chain IT facilities can currently be provided by:

1. An enterprise resource planning (ERP) package
2. Other specialist packages (e.g., customer relationship management [CRM])
3. Web-based systems
4. Other devices and systems providing a wide range of functionality such as bar code readers, radio frequency identification (RFID) tags, scanners, mobile communication devices, etc.

Supply-chain-related processes drive a large proportion of most businesses. They have a high impact on customer service, supplier relationships, and business costs. It is imperative that we continually examine IT developments to see how we can use them in our business processes. I do need to emphasize that we should drive implementation of new business processes, not IT. The availability of new IT is often the driver for change, but it is only the enabler. What matters is the ability of the new process to deliver more competitively.

An ERP package usually provides the databases and support transactions to run the business. Although packages vary and they may be customized, the functionality is broadly similar. Packages support sound, professional processes that perform well. However, they are unlikely to provide a great competitive advantage. Worse still, in some unfortunate cases, the IT system is seen as a monster—one that has to be fed endless streams of data, creating administrative work rather than reducing it. We know more than one company where good supply chain people drive the processes by approving POs, enter detailed importation data, and so forth. These people are well educated and highly motivated (they start that way at least) and this is a waste of their talent. It seems to me that the implementers of these systems have installed an IT facility, not a fully understood working process. Processes changes that remedy this type of problem are the use of pull systems rather than detailed computer schedules or using data collection devices for reading data such as serial numbers, country of origin, harmonized tariff code, and so forth. These process solutions often need combinations of IT. It is the ability to combine new devices, web-based systems and package databases that will be increasingly important.

A friend of mine runs a successful business developing web-based systems. They provide solutions that support business processes for retail sales and also services, for example insurance sales/claims/renewals. Their web-based systems are very cost-effective and provide a tailored solution for each business process. We can already see some ERP package suppliers taking a similar approach. Packaged solutions are likely to continue to provide most of the databases and basic transactions for the foreseeable future. They will be combined with web-based systems that support customers or a similar tailored set of needs, mobile devices for collecting and relaying information, and tags and radio devices for identification, information storage, and transmission. There is also much talk of *intelligent machines*. In my current experience, these are more like an extension of automation rather than intelligence. However, we are likely to see an increase in the number of ways the comparison of data and decision taking can be economically automated. This is already resulting in machines that can decide which parts they should make next, which parts should be picked and routed for transportation, which people are expected to attend meetings, and so forth.

Some examples of ways that technologies are being combined to support innovative processes come later in this section. First, here are some thoughts about getting packages to support the basic business needs well. There are fundamental issues that need to be correct, without which the whole basis of our processes, data integrity, and even the success of the business will be on shaky ground. The following sections describe the key IT packages relating to supply chain and give examples of the best practice issues for each system:

> Many of the ERP packages originated from MRP packages and their extension—particularly to include financial and sales functionality. The widening of their scope led to the ERP terminology. These packages have to do a good job for us, typically combining the following functions:
>
> - Financial ledgers for sales and purchases, and a general ledger, together with a wide range of facilities for management and financial reporting
> - Engineering systems such as bill of material (BOM) and engineering change control
> - Supply chain management systems for MRP, inventory recording, POs, and work orders
> - Sales systems for customer information and customer orders

Packages vary from the very simple costing a few thousand dollars to the most expensive costing well over $250 million including customization. Some suppliers provide little customization, some provide customization tools, and some are infinitely customizable using a combination of tools and unique programming.

Selection of an ERP package is a strategic decision. After some initial research, an RFQ is usually sent to a shortlist of potential vendors. It contains a specification of requirements, usually around business processes, and includes information about data and transaction volumes. It is important to specify the requirements of the processes, not how they should work. Vendors will then have sufficient freedom to propose new ways for the processes to operate. Critical aspects of the processes and important process and business goals should be included. The selection team evaluates the proposals and makes the selection and they agree on criteria for selection. The criteria should be shared with the potential vendors. This will help them to understand the relative importance of requirements. Vendors will respond with a proposal showing how their system would work and how much it would cost. This is likely to include presentations and demonstrations of their solutions. Key users are one of the most useful resources to judge how well these will work. They understand the real process issues, where there are difficulties, and can provide good input as long as they can be open-minded. Figure 6.4 shows some *best practice* issues for each of the elements under the ERP heading. We should stress that these are from the supply chain perspective and do not include issues from finance, engineering, or sales points of view. It should also be stressed that these fundamentals depend on a combination of the IT facilities and the process.

Purchase Ledger
Most system packages have been doing this for a long time and are very competent in this area. There are possible IT package issues in the areas of • International currency purchases. • Structuring of suppliers and parent companies. • Identification of purchases made by different product groups. • Report writer ease of use and capability to analyze spend in flexible ways. • Reporting of KPIs—purchase price variance (ppv), number of suppliers, spend by supplier and supplier group, spend by commodity type, payment on time. These are often on a *drilldown* basis. Other key issues for implementation and ongoing management are as follows: • Data must be *clean*, e.g., supplier records are not duplicated, company groups are identified. • Invoice clearance (three-way matching) and payment of suppliers on time—this is also likely to be related to process issues such as late booking of receipt transactions. • Supplier managers will need to maintain notes on a range of key issues for each supplier (a supplier relationship management [SRM] system). Links between the ledger and SRM may need to be thought about—for example, to update spend data or to restrict orders being raised on suppliers who are being *phased* out.
BOMs and Part Records
Bills of Material requirements vary widely depending on the complexity of the product. This may be one of the important areas for ERP package selection. • The ability of the ERP to manage variants and optional features may be an important issue in your business processes. Sometimes a *configurator* facility is required to enable product optional features and even dimensions to be specified to meet customer requirements. In some cases, this may be included in the ERP package, or it could be a stand-alone facility. The package must enable the product options and variants to be managed through a range of processes—costing, planning, manufacturing, assembly dispatch, and billing. • Engineering changes management (ECM) requires a combination of good IT support and good processes. Package capabilities will include effectivity dates, ability to use *old stock* first, historical change records, and management of engineering changes in MRP and kitting processes. An *Engineering Change Control Board* of some kind is likely to be an important part of the ECM process. Their needs for ECM management details such as spares provisioning need to be catered for. • Provision to access previous engineering versions sometimes needs to be made for service and maintenance. • Part descriptions are important to many business operations. For example, do we use the description "Screw-Cap Head DIA X LENGTH" or "Cap Head Screw DIA X LENGTH." Consistency in descriptions aids simple communication, reduces errors, and makes analysis easier. This is related to process rather than IT, and the source of descriptions is likely to be the computer-aided engineering (CAE) system. • Part classification is essential to aid many processes—i.e., is a part a standard fastener (screw, nut, bolt, washer...), a seal (*O Ring*, etc.), a machined item of a particular type (e.g., steel small turned), etc.? These need to suit the business, but need to be thought through carefully. Various standard codes can be used to classify parts, and you may choose to use one of these. • The source of part and BOM data can be a separate CAE system. The interface with the CAE system needs to support simple new and amended part number and BOM records and also the details of engineering change. Simple facilities to enable, for example, the transfer of new product BOMs from CAE to ERP BOM are useful. However, in my experience, adjustments to flags (e.g., phantoms) or even to the shape of the BOMs are likely to meet manufacturing and supply chain requirements. • In principle, the BOM should be owned by engineering, and they should be the *authority*. In best practice terms, BOMs should be simple, and only one version should be used by all operational areas of the business. Flatter BOMs (fewer levels) are simpler to understand. • BOMs need to be highly accurate if kitting and MRP-type functions are to use them. A 99% accuracy is needed if these functions are to run successfully. This can be measured by marshaling parts using the BOM and then assembling it—part shortages or excesses are likely to be BOM errors, and each one needs to be divided into the total number of parts on the marshaled list in order to calculate the percentage accuracy. Thus, one part short on a product with 100 parts at the final assembly level represents 99% accuracy. This is still likely to cause a significant problem, but is hopefully manageable if it is the exception.

Figure 6.4 ERP supply chain IT and process issues. *(Continued)*

Purchase Orders

Purchase orders can be used to:

- Document requirements to suppliers
- Enable receipts to be made against them
- Detail existing scheduled deliveries to the planning system (MRP)
- Enable *commitment reporting* (reporting of supplier commitments—in total, by supplier and commodity by time period)
- Enable *three-way* matching to check pricing and receipt quantities when *clearing* invoices

System IT facilities should enable these to be managed for a range of practical supplier arrangements:

- New POs, amendments and cancellations will be suggested by MRP. It is critical to establish how system parameters will enable dates and order quantities to be suggested in the practical and usable form you require. See Figure 5.2.
- Orders that are *firm*, *committed* (in quantity, but reschedulable), managed by *kanban* arrangements, or managed by *VMI* arrangements of various kinds.
- In some cases, *the Net Demand* from MRP is useful data to send to suppliers for planning purposes in support of *kanban* and *VMI* arrangements.
- The ability to manage purchase orders in foreign currencies (if absolutely necessary). This can be complex in the way exchange rates and timing are concerned.
- It is important to manage POs proactively—e.g., those that are overdue, 95% delivered, etc. need to be deleted, adjusted, or managed in some way. Good reports need to be made available to support this.

Do not use POs for low-value orders if avoidable (it is to administratively expensive). Use P Cards, one-stop shops, or VMI.

Inventory Recording

I find a *map* of the flow of inventory and its recording points very useful. Inventory recording transactions clearly need to be *leak-proof*, i.e., inventory must always be recorded correctly in quantity, value, and location until it is either sold or scrapped. System transactions need to be made as soon as possible after the physical inventory movement.

- A *map* of transaction types can help to understand this, for example,
  - o PO receipts could be made in Goods Inwards or on Collection from Supplier.
  - o Some transactions could be made automatically—for example, using scanners.
  - o Kitting transactions can issue all parts in one *hit*, but how are phantoms, items managed by VMI, or Kanban-type arrangements managed?
  - o Where are the receipts into *holding locations* such as *Goods Inwards* and how do transactions move them into their recorded locations?
- Multiple warehouse locations need to be managed, and if required, enquiries and requests for transfers of inventory supported.
- Are locations fixed or variable, and how are they to be managed?
- Does the system tell us in which location to put incoming goods? (The layout of warehouses and inventory locations must be logical and work in conjunction with the picking list-type functions to enable accurate and fast location of parts and fast/easy access to frequent picks/heavy items.)
- How does the system manage ageing of parts—FIFO, expiry date, etc.?
- Arrangements for *allocating* inventory, for example, to customers, orders, kitting, or product build purposes need to be clear. There need to be clear conditions under which inventory can be allocated and when it cannot. Some types of inventory might not be allocatable to customer orders (not saleable parts), and these rules should be enforced by the IT system.
- Cycle counting facilities need to be included in the IT system and meet the practical needs, i.e., they need to enable planned cycle count requests to be raised in the best way for you—e.g., by sets of locations. Accuracy levels close to 100% are now common; less than about 96% in my view indicates that fundamental problems exist. Quality of the warehouse locations should also be facilitated by reporting of quality issues such as safety incidents and 5S assessments—it is useful if this is by area or by people/team responsibility.
- Dangerous goods or items that require special storage such as chemicals, radioactive, electronic items, etc. should be clearly identifiable, and access to the correct storage requirements and regulations should be clearly referenced and accessible.

Figure 6.4 (Continued) ERP supply chain IT and process issues. *(Continued)*

Work Orders
Work orders are used to • Check availability and reserve parts for assembly • Produce picking lists for the warehouse • Enable all parts to be de-stocked in one transaction • New Work Orders, amendments and cancellations will be suggested by MRP. It is critical to establish how system parameters will enable these to be suggested in the practical and usable form you require. See Figure 5.2. Bill of material flags need to be set carefully to cause work orders and picking lists to manage and list parts in line with what you want to happen physically, e.g., Phantom flags to ignore BOM levels you don't want to transact, Shop Stock or Floor Stock flag to indicate which items are already available and should not be included in picking or marshaling lists (e.g., because they are serviced directly to point of use by the supplier under *VMI* or *kanban* arrangements). Work orders need to be managed proactively. It is good practice not to raise the orders too early and close them off properly—no old ones which were started but never completed for example.
Sales and Operations Planning (S&OP—See Separate Section)
It is important to stress that Sales and Operations Planning is a *process*, not an *IT system*. However, the amount of data that need to be gathered, processed, and then fed back into the planning system can be quite large. The IT support system is therefore important. Many businesses run S&OP successfully using spreadsheets (albeit in a fairly sophisticated manner with formulae and links, data transfer to and from the package databases, strict version control, etc.). This enables those running the process to format the data in a highly customized way. For example, they may have lines of data showing forecast and actual demand, current inventory and production plans, and the projected balance between the two. They can then adjust the production plans, see the result in inventory or customer order-book terms, and also see the financial values such as projected sales. If you decide to do this, even for the short term, it is important to *fix* the format of the spreadsheet and automate the transfer of data from and to the main system. Data from the main IT system will include 1. Current sales orders 2. Forecast sales by product 3. Current inventory of saleable products 4. Current manufacturing/purchasing plan The agreed manufacturing plan needs to be transferred back into the main system for implementation—usually to run MRP. S&OP facilities provided as part of packages vary in sophistication. Some provide what is little more than a single-level BOM breakdown, some provide similar facilities to the spreadsheet methods described above, and some have much more complex facilities—for example, for dealing with product variants. Some other issues regarding package solutions are as follows: • The parts or products that S&OP uses are of course the saleable items. Where there are product variants, some kind of representative product or planning bill needs to be used. Obviously these need to be coordinated across sales, forecasting, and manufacturing. There will need to be some kind of BOM processing to convert, for example, sales orders for specific products (with options) into the right items for S&OP purposes. • Management typically needs to see S&OP data in financial terms, but product units are usually used by those managing the process. A method for conversion between units and financial values is needed. Financial values are usually based on selling prices. However, maintenance of up-to-date prices, customer discounts, volume discounts, and foreign currency sales often makes this difficult. All the methods I tried resulted in some level of errors that need to be recognized and managed. • Some businesses need to commit to dates when product sales to customers will become available from production. This needs to take into account inventory, existing sales orders, and planned production/supply. Some packages provide this functionality, which needs to be calculated period by period, starting with the most distant future period. • It may be useful to use a spreadsheet S&OP as a prototype. However, if S&OP support can be provided as described using spreadsheets, there is no reason why they cannot be provided in the main system. Data transfers are not then needed and version control becomes easier. However, most of the key process issues remain. One issue often overshadows the IT process—that of senior management involvement in the process. They need to have confidence in the assessment of the balance of supply with demand and the taking of risk decisions—for example, in committing to future purchases or manufacturing.

Figure 6.4 (Continued) ERP supply chain IT and process issues. *(Continued)*

Material Requirements Planning (MRP)
Most MRP systems are broadly similar. The systems can run periodically (e.g., twice per week) or continuously. Even planning parameters are similar (e.g., batch quantity/quantity multiples, period order quantities, etc.) The precise way in which orders are suggested in the MRP run does vary. It is essential to understand this and I recommend experimenting with the system in some kind of test environment. Set up situations that you need to understand, for example, • Projected shortages within lead-time (packages will not generally raise orders within lead-time). • Projected shortages of small (insignificant) amounts (some packages have *sensitivity adjustments* to enable these to be ignored). • Projected shortages at earlier dates than existing orders—how are existing orders rescheduled, or new orders raised—and for what quantities? • Engineering changes set up by effectivity date—how are any orders for old parts treated (both before and after the effectivity date), and how are new orders raised? • If there is inventory of a *phantom* part, is it used up before part requirements are planned at the lower level? This facility is sometimes used to implement engineering changes where inventory of the old part is to be used before starting to use the new part. • Projected shortages at earlier dates than existing fixed orders exist—are new orders raised in the earlier time period? • Use a combination of batch size parameters to investigate how they work together. Keeping planning parameters simple helps in understanding what will happen under a range of planning circumstances. Ultimately it makes management and control of a stable MRP easier. Good S&OP planning will also help to prevent MRP from doing the wrong thing. S&OP rules should prevent plans being changed within less time than can be achieved. The plan that is the input to MRP should therefore be achievable. New demands that are unachievable will of course arise when, for example, parts are scrapped, or suppliers report they cannot meet the required due dates. The behavior of the MRP system under this type of circumstance needs to be tested and fully understood.
Sales Order Management
Sales orders are typically received via a web-based facility, on which a customer has selected the product and product options required, or by a sales person using a similar facility. The web facility may be separate from the main system (ERP or another package). However, the links with other functionality need to work quickly and accurately: • The availability or delivery date of the product required. This facility could range from a simple inventory reservation system to one that allocated product from a future available plan. If there is any system that adds forecast to actual sales in order to assess total demand, it is essential for incoming orders to *consume* the forecast. This means that for every product order added, one must be removed from the forecast. Otherwise, some demand will be *double counted.* • Optional product features must be defined in the same way as the forecasting and planning functions. Otherwise, they will not be able to be managed consecutively. • Existing customers and their details need to be recognized, and new customers need their details collecting, verifying, and registering. • Discounts may need to be applied based on a number of mechanisms—customer, order size, region, etc. • Most sales organizations have targets by regions or customers or sales people. The sales order management system must be able to attribute sales to one or more of these categories. • It will also be useful to collect data alongside the customer orders, which correlates with the forecasting approach, i.e., geographic regions, types of customer, and reasons for purchase (e.g., new build/renovation) under which sales can be grouped in order to compare with the forecast numbers.

Figure 6.4 (Continued) ERP supply chain IT and process issues.

The management of customization, which is part of most implementations, is a difficult job for the IT department. They need to: involve users, manage the package provider, produce an effective business processes, work within budget, and satisfy senior management. This is an important management role and should not necessarily be carried out by the best technical expert. The project manager may be an IT professional or is as likely to be a key user, manager, or even an external resource of some kind. They need to understand IT solutions, but more importantly, be a good manager of people and resources and have a good understanding of what process performance issues are important to the business.

6.2.1 Other Packages

The following are also common packages used in the supply chain.

6.2.1.1 Forecasting The forecasting package used is often not part of the ERP system. Good forecasting packages are available that use a range of techniques and are suitable for most businesses. The best ones enable a range of techniques to be used together and usually offer a *best fit* option. Generally, there are three approaches to forecasting:

- *Prospect led*, where salespeople enter their sales prospects from a range of existing or potential customers. Required dates and percentage likelihood of making the sale (see Section 3.1 and Figure 3.1).
- *Correlated*, relating forecast sales to some known indicators such as current rate of house building or economic data.
- *Historic*, usually by product, basing future sales on past history. This can be quite sophisticated, including seasonality, trends, and alpha factors to indicate current economic trends or other influences.

A client has a *best practice* process in which *prospect-led* forecasts are entered for large customers that are then compared with *historic* forecasts made by the forecasting system. Significant differences are then examined, which has the effect of questioning the *prospects* and adding more insightful data to the *historic* forecast. Even using this very good approach, in this global retail business, individual product forecasting can be 50% out over the 3-month manufacturing process time. As the saying goes, forecasts are either lucky or wrong.

Some ERP packages do include forecasting facilities similar to the independent packages.

A recent trend in businesses with complex variant features is to forecast at lower levels of the BOM by analyzing past demand and current trends. This can be at a lower level of BOM than that that sold to the customer. These appear to be useful if the number of product variants is very high, which make the planning bill approach too difficult, or where there are a lot of spares or maintenance demand intermixed with new build demand.

6.2.1.2 Warehouse Management Systems (WMSs) WMSs are inventory management modules in all ERP packages. These are generally quite basic, but work well. They provide real-time inventory recording, inventory location, picking list facilities, cycle counting, and reporting support.

WMS packages are used when there is no ERP package or when additional facilities are required. An example of this could be for automated warehousing.

A WMS includes all the basic warehouse facilities of inventory and location recording, picking lists, cycle counting, and reporting (provided by modules in ERP packages) but additionally could be used to support

- Stock replenishment from suppliers
- Automatic creation of POs
- Cross-referencing part numbers from different suppliers
- Inbound freight management
- Fully automated, paperless receipt of goods
- Automated put away of goods to optimized location(s)
- Automated picking of goods
- Direct picking to shipping carton
- Mixed picking of customer orders from location for collation at packing
- Fill rate optimization
- Shipping documentation/data
- Notification of dispatch to customer
- Handling of product returns from customers (interfaced with the financial system)

Most WMS implementations I have seen include integration with some form of automation ranging from automated handling shelving systems to automated fork-lift trucks and pallet storage. These systems and the equipment that they interface with are becoming a sophisticated and specialist area.

6.2.1.3 Distribution Network Planning Systems Distribution network planning systems are package systems, and a range of them are available to optimize distribution. Some packages work on a strategic basis, optimizing the positioning of warehouses and routing of goods both into the warehouses and onward to customers. They can use a range of transportation types—road, rail, air, and sea. Some packages work on a tactical basis to optimize the loading of deliveries to trucks with optimized delivery routes. Some packages do both jobs. Cross-docking* is often used to transfer goods from trunking vehicles to more local delivery vehicles. This can also be managed by most packages.

One client makes and delivers fresh produce to shops all over the United Kingdom every day. The product is made in several factories across the county and is collected

* Cross-docking describes arrangements for receiving goods from one set of vehicles and re-sorting them as they cross the dock to be dispatched again in new loads into a new distribution network. Typical examples would be the receipt of goods from trunking vehicles into distribution points where they are re-sorted for dispatch to customers in loads on more local transportation. Supplier deliveries of perishable goods are often received to cross-docking operations from which the retailer re-sorts them for delivery to stores. These operations are often very time-specific because the final delivery may be to a store for which they have been given specific delivery time slots. Some operations trunk nationally overnight and cross dock and distribute all across the United Kingdom by the end of the morning.

and trunked to distribution warehouses. The product is then cross docked, selecting product and making up route loads destined for retail outlets. The distribution warehouses have been strategically positioned based on the system's analysis and the package is also used to optimize routing on a daily basis. This system will even rework distribution routes when a vehicle breaks down.

A typical strategic approach would be to model the expected end delivery requirements. This is likely to be a list of customer locations, delivery dates, and package sizes and weights. The sources of data are usually historical records from sales systems, delivery manifests, and a product database. The historical delivery requirements may need to be adjusted or updated with growth, changing delivery patterns, or changes in delivery areas. The network planning software then compares the delivery requirements with the position of supplying locations and existing and potential warehouse locations. The system determines the optimum loads, routes, and transportation types for incoming goods and distribution. The system will take into account the customer service levels required. Changes to warehouse location can be experimented with and the effect on total cost and customer service level will be modeled. Costs of warehousing in different locations will be taken into account as well as the cost of transportation. These systems are not restricted by country borders. In fact it is common for this type of system to be used to optimize European distribution. These systems are very impressive and can take into account a wide range of issues such as goods-vehicle weekend driving restrictions in some countries. The process of optimization with these systems is iterative—knowledgeable people examining the modeled results and making adjustments to check the effect of, say a new warehouse location.

6.2.1.4 Customer Relationship Management (CRM) CRM systems enable sales to know all they possibly can about their customers. The normal customer record types of data (names, addresses, contact details, company profile, past order history, and forecasts) are supplemented by almost anything that could be useful, such as names of key people (and some notes about them), meeting notes, discussions about potential business, conversations about the customer held (or even overheard) with third parties, news articles, or stock exchange results. This can be a sophisticated area and continues to develop. Some systems are even collecting information from social media.

This is not a supply chain system. However, for supply chain people, the challenge is the amount that the salesperson may know about us and our business. It can be embarrassing if they know more about our business than we do, but even worse if it enables them to do better than they should in a negotiation.

Those who are responsible for the supplier relationship or who are negotiating also need to know about the supplier's affairs in the same detail.

6.2.1.5 Supplier Relationship Management A supplier relationship management (SRM) system is like a mirror image of the CRM Section 6.2.1.4. It should enable supply chain people, in particular those in purchasing, to know equally as much about their suppliers as their suppliers know about them.

The system should store the supplier's company details (although these should be mirrored rather than duplicated on the purchase ledger), addresses, people's names and positions, financial details and history, notes about the business, and so forth. (Note that it is important to arrange who updates supplier details on the purchase ledger and on SRM.)

The SRM system should also store details of events such as meetings, highlighting actions to be taken.

Plans for improvements, KPIs, current performance, and targets should also be stored where appropriate.

6.2.1.6 Computer-Aided Engineering (CAE) From the engineering point of view, the CAE system supports the whole of the design process. This can include component and product designs, libraries of parts, three-dimensional (3-D) design tools including many support facilities such as stress analysis and tolerancing, BOM and engineering change management, product costing, and many other facilities. Many industries, aerospace for example, have adopted a single package for use throughout the industry, with the obvious advantages of standardization and transferability of designs.

From the supply chain point of view, CAE is important as the interface to technical product data and as the driver of good practice on standards. The following three aspects are important for the supply chain:

- Simple transfer of new and updated product data to supply chain databases (mainly MRP for planning, material movement, sourcing, and product costing)
- Ease of transfer of design data to supplier systems (e.g., 3-D modeling of component designs can be transferred to machinist or molder suppliers and used to program manufacturing operations)
- Support of best practice for standard format descriptions, use of common parts, materials, and processes (e.g., ease of searching for *preferred parts*/promoting part rationalization for standard materials that are already in use)

Part numbering scheme should be appropriate. Most systems now use insignificant numbering (i.e., the part number does not include any significant coding by which parts types can be identified).

A naming convention of some sort is needed. Apart from the clarity of understanding this brings, regrettably, we often need to resort to searching and analyzing, using the part description field (see BOM notes above).

Dangerous goods need to be classified and reference to storage requirements and other standards made easily accessible.

6.2.1.7 Project Management Packages Project management package systems are based on the programming evaluation and review technique (PERT) and are used to both plan and manage projects. There is a wide range of packages available. Project management functionality is also a standard feature of some ERP systems. The activities that have to be completed for the project are entered into the system together with information about the duration of the activity, which other activities need to be completed before this one can start, any overriding dates regarding when the activity can start or finish, and what resources and costs will be required. The system arranges all the activities in a logical sequence and works out the total duration of the project. Critical paths are worked out—sequences of activity that determine the finish time for the project (other paths of activities do not contain as much work and there is slack time that can enable the start and finish times for some activities to float).

Critical to most projects are contractor and supplier activities, materials, parts, and large sections of plant. These are clearly supply chain management issues and they can be managed and interfaced with the project management software in a similar way as with MRP (i.e., specific activities can be set up for services or items to be supplied). This can prompt for orders or schedules to be raised and managed and for the costs to be collected. Slippage in the time scales of activities are fed into the system, which will enable it to replan all activities and the project time scales if necessary.

Getting the fundamental process and IT support right is essential to basic professional operation.

* * *

6.3 IT to Support Competition-Beating Processes

We need to look beyond the fundamentals and package solutions to achieve true competitive advantage. This is most likely achieved by innovative application of a range of IT along business processes. The range of IT available is expanding all the time, and include

- Web-based applications, which are easily customized to allow customers to view and order online, show pictures and sizes, and send messages to other devices
- Mobile applications, used to send people messages and updates, collect data, or to relay it to people or machines
- Identification devices, such as RFID tags attached to goods, raw materials, components, products, or machines to carry their identity, traceability, technical details, position, or the current temperature, current pressure, and humidity they are experiencing
- Scanning devices, which record the movement of parts and materials or to direct them to loading bays or vehicles
- Massive ability to store, access, and process data
- Intelligent machines, which collect and analyze data and take appropriate action

Differentiating our service in a way that is seen directly by the customer has a high potential impact. Web-based advertising and ordering, sales support, delivery information, and tracking systems are obvious examples. Businesses offering web-based sales and direct delivery implemented new IT supported processes some time ago. They are a good example of the use of combined, innovative IT. Customers now expect the following facilities to be standard:

- Web-based sales enabling
  - Direct customer access online, including using mobile devices
  - Presentation of products in pictures and specification
  - Checking of inventory availability
  - Can be interactive and answer questions/take customers through a logical route to purchase
  - Frequently updated with new designs, prices, and offers
  - Existing customers are recognized, leading to advanced customer contact and customization to known preferences
  - Notification that their purchase has been dispatched, confirming their preferred delivery option and time, and providing tracking details
  - Up-to-date delivery tracking information, driver details, delivery vehicle and current location, schedule of other deliveries to be made, and latest delivery time estimate
  - Product configuration

A typical current process for customer purchase and delivery is shown in Figure 6.5.

Innovative combination of IT along processes produces competition-beating processes.

* * *

6.3.1 Other Examples of Good Uses of IT

6.3.1.1 Vendor Managed Inventory (VMI) There are many approaches to enable vendors to respond to customer demand and deliver matching inventory. VMI is used by a high-volume manufacturer of snack foods that are sold across Europe. A special piece of packaging is supplied from a factory some distance away (in another country) and is able to supply on a VMI basis, as shown in Figure 6.6.

Demand, supply, and inventory balance information is seen by both customer and supplier. Changes are updated quickly, enabling rapid response when the projected balance of inventory at the customer premises is out of the agreed tolerance. The system is startlingly simple and yet highly effective.

Another approach, an even simpler one, is at an aircraft maker's plant. A type of strut is supplied to the European factory from the supplier in the United States. Shadow lines are painted on the floor of the supplier's factory in the position where the

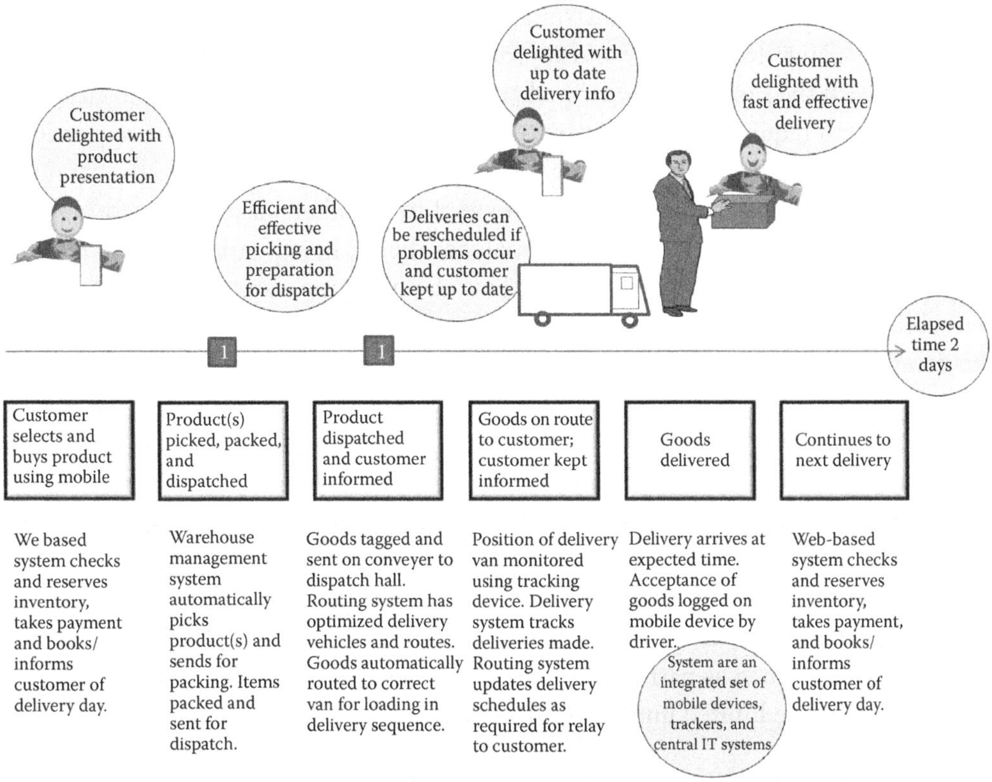

Figure 6.5 Delivery of goods to customer using a range of IT devices.

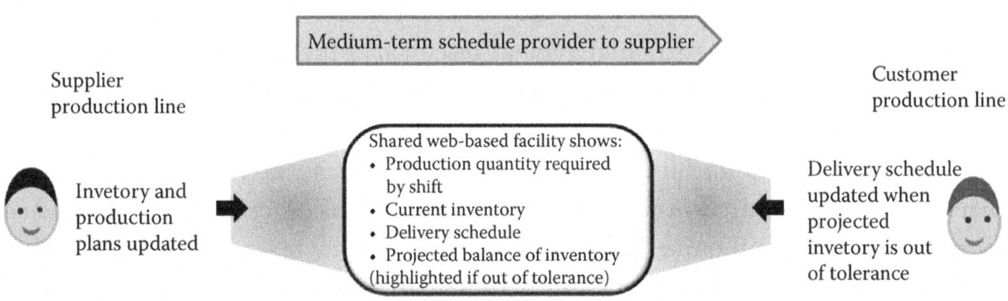

Figure 6.6 IT support for VMI supply.

target quantity of struts should be. When one is used, the shadow lines of the component are visible rather than the component itself. A webcam (I have counted this as IT!) is continuously viewing the shadow lines area and is monitored by the supplier over 6000 miles away. The supplier immediately dispatches replenishment inventory by airfreight.

Car makers need frequent deliveries of assemblies, but they are complicated by the variant requirements of each vehicle to be produced. Most use some form of sequenced

VMI. An example of this is for instrument panels delivered from a factory close to a car assembly plant. The quantity and approximate mix of car variants has already been agreed on with the supplier. The car maker schedules each car to be made to customer order. This will require specific features on the instrument panel and the sequence in which these will be built is accessed by the supplier using a special link to the car maker's system. The instrument panels are provided in mobile racks, sequenced in order for the robot to pick the next panel that matches the car being built. Each panel has a simple bar code containing the panel serial number that provides technical details and traceability. The bar code is read by the robot as it is positioned for car assembly and the details are added to the records for the car.

6.3.1.2 Industry 4.0 The term industry 4.0 is currently also being used to describe a combination of data, communication, and intelligent machines, but in a more industrial context. An example of this would be machine tools that are aware of the schedule of parts to be manufactured, can identify and select the raw material required because it has a RFID tag, can make decisions about the most effective job to work, load the programs to machine the part, and reattach a RFID tag to the completed part that includes data such as part traceability. This is also truly impressive. However, the applications will be limited until costs are lower.

6.3.1.3 Point-of-Sales Data We were working in a biscuit factory on a Saturday morning, part of a very large European company, when the inventory systems started to go into panic mode. Point-of-sale (POS) data had been used for some time by the supplier to monitor and provide early warning of sales patterns in retailer stores. On this occasion the sales of one particular type of biscuit had rocketed and replenishment inventories in warehouses were inadequate. A special offer had been arranged with retailers for the weekend, but nobody had told the supply chain or manufacturing. These systems are very useful even in in normal circumstances and they enable suppliers to respond rapidly to end-customer demand. Decision making about unusual demand patterns is still quite difficult, of course. Is the increase in demand likely to continue, or will demand now be lower because everyone now has a cupboard full of biscuits at home? I haven't seen the IT system that can investigate and take this kind of decision, so the process is likely to include a planner of some description who will hopefully make the right call.

These are a few examples of good supply chain processes that depend on the exploitation of a combination of IT solutions combined with intelligent humans and business processes. Off-the-shelf package solutions can form the platform for data storage and for specific processes such as MRP. However, these will only achieve me-too levels of competitiveness on their own.

A range of IT now needs to be combined to produce innovative processes with a differential level of performance.

***

The development of new and innovative processes that take advantage of new IT capabilities requires a range of skills. The important ones are

- Knowledge of the current process, what it does well, and where it needs to be improved
- Knowledge of the competitive impact of the process, its costs, response, and quality issues on competitiveness
- Knowledge of what a wide range possibilities new IT can achieve

These need to be combined in a way that investigates the possible solutions and develops a process that will exceed customer and business expectations. The result could be a new process design, well-chosen IT solutions, a costed implementation plan, and estimated business benefits. These need to match the competitive objectives of the business.

The ERP system is the central supporting system for most businesses. The ERP package has a very practical impact on customers, suppliers, and almost all employees. A large proportion of what the ERP system supports are supply chain processes. We therefore need to make sure our processes and systems support deliver the right results. This is a big task. There are a lot of high-level issues to get right and yet as usual *the devil is in the details.* The performance of the process is the crucial issue, not the system. If you are involved in implementing or reviewing processes and systems, you may find it useful to draw a process grid similar to the one shown in Figure 6.7. The key process stages, process and performance issues, and system support that could differentiate will depend on your business. It is important to think of a wider range of IT solutions than just the ERP package. Think of all the other technologies that can be interfaced with the ERP solution and how competitive advantage could be supported. In the supply chain, competitive advantage will generally be by

1. Delivering at lower total cost (this can include a range of indirect costs, including supply chain administration)
2. Delivering with faster response
3. Delivering at higher quality (this can be product quality, service quality, or even the integrity of the supply chain)

Figure 6.7 shows an example grid of some key supply chain process stages, the process performance issues, and possible system functionality that might enable us to differentiate ourselves from our competitors.

The example is of the main Supply Chain process in a typical product manufacturing business.

Process stages								
Sales forecating	Sales and operations planning	Sales order processing	Inventory management	Supply chain planning	Purchase order management	Internal work scheduling	Assembly scheduling and control	Shipping
Process/performance issues								
• Forecast accuracy	• Understanding of planning decision consequences	• Fast response to customer • Accurate availability information	• Movements transacted quickly • Data accuracy	• Fast response to change • Stable • Changes easy to implement	• Fast and accurate alignment of demands with supplier capability	• Short lead times • High utilization	• High satisfaction of customer demand • Short lead times	•Fast customs clearance • Lowest cost achieved for customer service required • Accurate shipping data
Possible system functionality that might differentiate								
• Improved analytical techniques and combinations of Historical trend, correlated and prospected data	• Modeling to access impact of what-if scenarios	• Web-based product selection and configura-tion/ ordering	• Automated data entry • RFID tags	• Linked to key supplier systems or web-based facilities e.g., VMI data linked	• Linked to key supplier systems or web-based facilities e.g., VMI data linked	• Local cell/ shop floor detailed planning linked to central system	• Local cell/ shop floor detailed planning linked to central system • + llinks to customers	• RIFD tag carries copy of product and shipping data

Figure 6.7 Process/issues/systems differentiator grid.

Here are some other very important processes relating to the supply chain:

- New product development
    - Easy access to, and use of, current parts and standards
    - Fast process for obtaining prototype parts
- Engineering change
    - Impact on existing inventory and provision of new parts easy to assess
    - Ability to communicate changes internally to customers/users and suppliers
    - Ability to implement phased change in planning system
- Customer cash collection
    - Fast and accurate transaction processing
    - Automated cash collection
    - Exception reporting
- Invoice payment
    - Fast approvals process
    - High first-time pass on three-way match
    - Low data errors on values, dates, and supplier details

Identify where the important performance issues sit along the process and how IT could practically help us differentiate from our competitors.

* * *

There is a danger that some new IT is used *just because we can*. On the other hand, unless somebody does take this approach, many advances will not be made.

Figure 6.8 shows a range of approaches to IT work that can be taken. In most cases more than one of these needs to be underway at the same time.

1. Support and maintenance of existing applications is needed unless they are about to be replaced. IT resources and some form of user team are needed to prioritize and keep applications working.
2. Continuous improvement of existing applications can respond to user feedback, KPIs, and Lean work such as process mapping. A user team is needed to prioritize the value of improvements and IT resource is needed to do the work and ensure other developments are not about to overtake the improvements. It is easy to underestimate the value of this work when make existing applications more productive.
3. New technologies can often be linked into existing IT systems to improve the effectiveness of the process: RFID tags for example, can reduce the amount of data entry required and reduce process time and error rates. Scanners can trigger automatic transactions to record the movement of goods. Mobile devices can enable data inquiries and transactions to be made remotely. The ratio of benefits gained compared to costs and risks from this type of work is likely to be good.
4. Web-based systems are common in applications such as direct sales. The user experience provided by applications that are tailored for example, to recognize customers, provide special discounts, and display pictures of goods for sale improves competitiveness. The applications can be tailored to precise

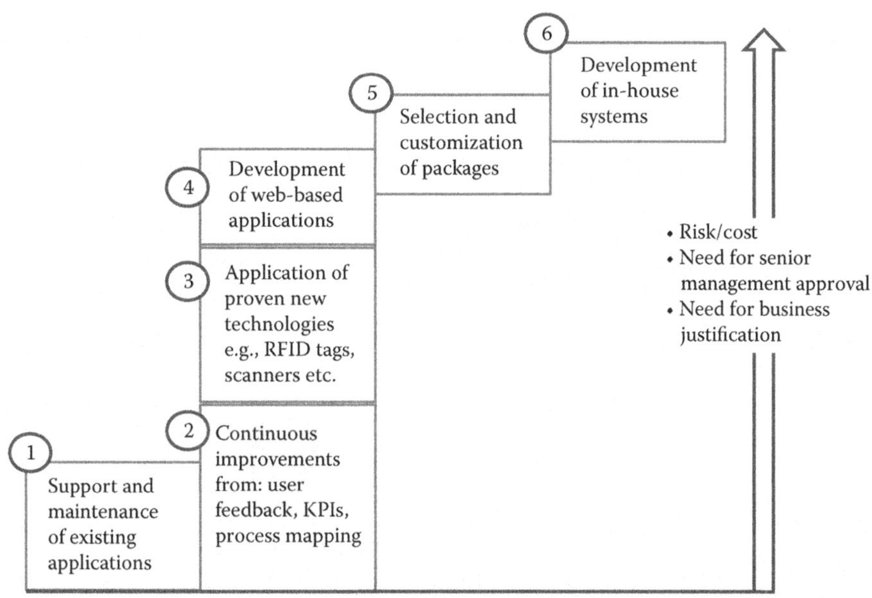

Figure 6.8 Program positioning matrix. (Adapted from the Programme Portfolio Matrix from Cranfield University School of Management.)

requirements and are relatively low in cost. It seems inevitable that this type of application will replace many of the functions currently provided by the transactions of software packages. Some packages are themselves changing toward providing customizable databases with web-based transactions.

5. Packages will continue to provide the basis of IT systems for many supply chain and business systems for some time. I have already discussed the process for selection of packages. They are usually quite costly and senior management approval and steering is required.
6. With the exception of web-based systems (which I have categorized separately) few supply chain systems are bespoke these days. The costs of development and risk of making mistakes are high. Few applications can justify bespoke software. Possible exceptions could be supply chain specialist businesses, such as global shipping companies, for example. However, even in these cases, web-based systems are more likely to be able to deliver bespoke requirements at a much lower price.

It is likely there will be a trend toward work of type 3 and 4 because these offer more flexible solutions at a lower cost.

The right mix of people must be involved in deciding how IT can best be implemented to deliver something that differentiates us from our competitors. Will the developments bring an outstanding new level of customer service, faster response, higher quality, or lower costs?

In one form or another, supply chain processes represent the major part of what IT systems support in most businesses that make or sell a physical product. The impact of

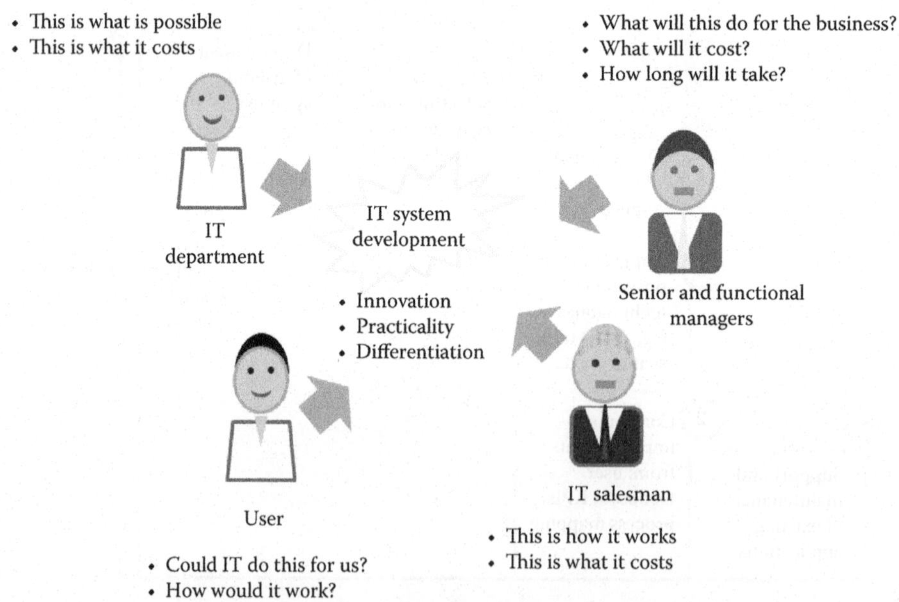

Figure 6.9 Who can contribute to the development of innovative IT development roles?

these processes and the IT that supports them has a direct impact on competitiveness. Therefore, substantial effort needs to be made to develop them to be as good and as effective as possible.

The IT profession has a serious role to play in explaining what is possible. In most cases, they need to publicize what new technology can do, perhaps with some blue sky thinking about how it could be applied. Senior management and managers responsible for key processes have a role to play in developing the *vision* of what could be done and what business impact it could have. However, the work of deciding how the system supports the business process needs to be heavily influenced by the people who actually manage and do the work. There are obvious inputs from people in different roles, perhaps something like what is shown in Figure 6.9. The people involved also clearly need to have a range of characteristics to enable solutions to be both bold and innovative, but also practical and affordable.

IT support and supply chain processes need to be developed by teams combining technical skills, business vision, and firsthand knowledge of the processes.

* * *

7

Supply Chain People and Organizations

7.1 Communication Skills and Quality

Communication is the most important skill for supply chain people. Supply chain people optimize the supply of goods and services from suppliers through our businesses to the customer. In this work they interface with more different departments and suppliers than any other function. It is also clear that good supply chain people do also have excellent numeracy and logical thinking ability as well as great communication. This applies right from the head of the supply chain to vendor schedulers, buyers, and warehouse managers.

The supply chain is a demanding work area requiring high levels of skill in diverse areas:

- Ability to communicate accurately and effectively, often including technical or quality data, at levels from office and shop floor to the boardroom and with suppliers
- Knowledge of relationships between businesses (commercial, contractual, or technical)
- Knowledge of the business they work in and the products and services they manage
- Sufficient technical understanding of products or services they buy
- Capability to understand, manage, and develop planning and risk management processes
- Knowledge of specialist areas such as trade compliance, customs clearance, and shipping
- Ability to work as part of a team and to have the drive, tenacity, and versatility to get things done

In addition, because the supply chain is not a long-established function like finance, in many organizations people are often struggling to achieve legitimacy. Other departments, such as engineering, operations, marketing, and project management, all think they can source or buy things themselves. Practical issues often do not go to plan in the supply chain and criticism is often heaped on the organization in fine style. For this reason, supply chain people have to perform at an even higher standard of effectiveness and professionalism to be accepted for their true value into many teams.

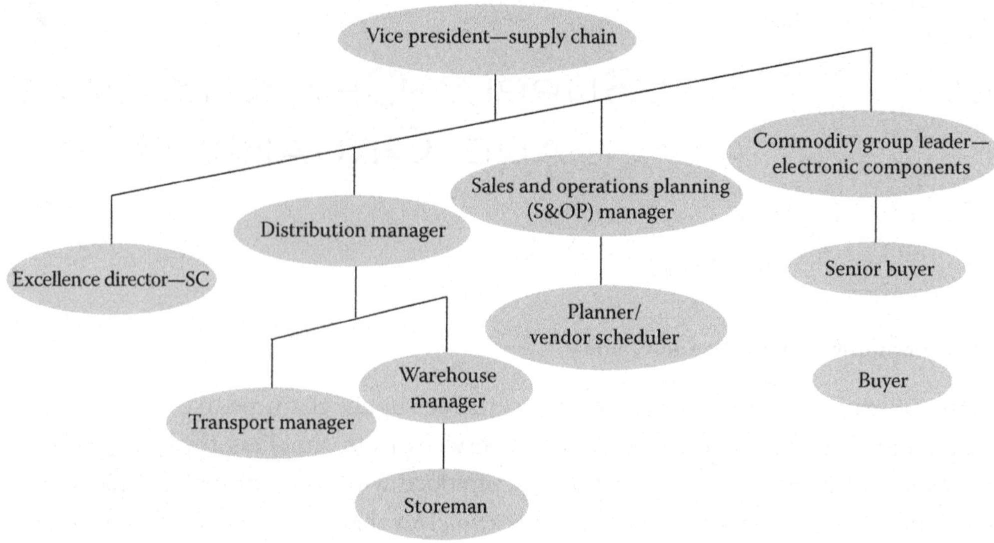

Figure 7.1 Supply chain people and organization.

In order to review the wide range of types of people and skills needed in a supply chain organization, we show role definitions for some typical positions in a hypothetical structure in Figure 7.1.

7.1.1 Vice President, Supply Chain

The most senior supply chain person needs to both represent the supply chain at board level and develop the capability of the organization and supply base. In many cases, significant improvements are needed by the business. The vice president (VP) is often selected to improve particular areas that the business judges to be a priority. These could be the capability of the supply chain organization, the shape and performance of the supply base, globalization of supply, reduction in cost of supply, or in the cost of administration or distribution, for example. The ability to communicate, to lead, and inspire teams across the range of supply chain tasks can be particularly challenging.

7.2 Organization Structures, Teams, and People

If we were starting an important project involving a range of diverse people who need to cooperate fairly closely in doing their work, we would *not* organize them under different managers and locate them well away from each other. Put like this it sounds perverse, but in many businesses, changing the way people are organized could bring improvements in customer focus and efficiency at lower cost. Many businesses are starting to think about new organization structures and colocation, moving away from the old departmental type of structure to organize their people around the work or processes.

Locating people together in the same area would certainly help with communication. Even if there are too many to locate in an office or the floor of a building, we could locate them in subgroups or teams based on the flow of work rather than their functional skills. Figure 7.2 indicates how unnecessary simple e-mails might be between people who sit close by.

Reporting structures also need careful thought. We could organize reporting to team leaders focused on getting the job done with customer satisfaction and profitability as key drivers. This is likely to be efficient and effective, and possibly quite motivating to the team. However, standards for areas such as fiscal control, engineering, or health and safety might be compromised. A functional departmental structure might maximize adherence to standards but might also be less efficient. Time scales are often extended because communication is more remote, objectives and performance measures may also become less customer focused, and in some cases confusing or even contradictory. *Matrix structures* try to address this problem, and succeed with varying degrees of success.

Manufacturing was among the first to realize the potential of reorganizing resources. In the 1970s and 1980s machine shops were reorganized in what was called *group technology*. Previously, there had been separate sections of lathes, mills, drills, and grinders in a machine shop, presses and bending machines in a sheet metal shop, and various welding machines in the weld shop. Group technology rearranged these machines into groups, each with a mixture of machines capable of making a particular family of parts.

The effect of this is shown in Figure 7.3. On the left is an old functional machine shop. On the right is group technology from the 1980s. The different styles of arrows represent types of part entering the shops and progressing to the types of machine needed to produce their shape. A much simpler route for parts to be made is evident for group technology. Benefits include shorter lead time (although the machining time may be the same), better part knowledge/familiarity, and easier fault diagnosis and correction.

Figure 7.2 Colocation improves communication.

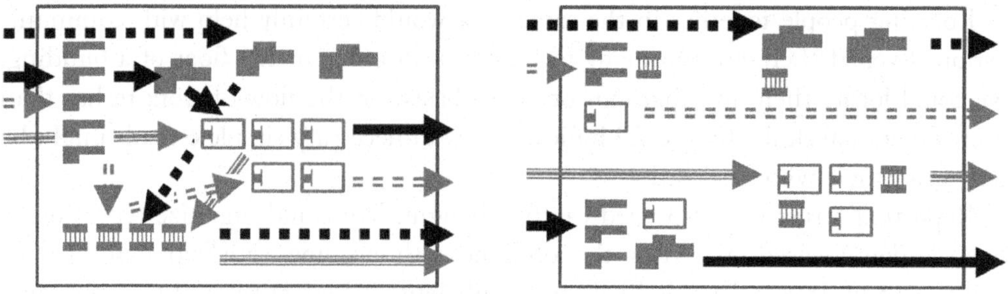

Figure 7.3 Conventional and cell-based workflow. (Left) Old functional machine shop. (Right) Group technology machine shop from the 1980s.

Cellular manufacture followed group technology quite quickly in the late 1980s and 1990s. In simple terms, this added the people dimension to the thinking. Manufacturing cells contain the people needed to run it. This includes not only machinists, but for example, people to do engineering support, costing work, and material scheduling.

There can be a number of manufacturing cells, for example, in a business making one product range. The cells could make a range of machined, fabricated, and electronic and electrical parts and subassemblies. In this case the cells need to be coordinated to flow parts and subassemblies to final product assembly.

In many cases product manufacturing cells have responsibility for tactical customer contact, supplier scheduling, product costing, quality improvement, and people management. In product manufacturing or maintenance cells this can improve the accuracy and timeliness of information flow with suppliers and customers because the people responsible are colocated and have factual and up-to-date information. The sense of ownership and increased motivation is also significant. The term "mini business" has been used for this type of cell because this accurately describes its functions and responsibilities.

Although tactical responsibility for engineering, customer service, supplier scheduling, costing, and quality is given to the manufacturing cell or mini business, these functions need to be coordinated across the business. Suppliers, for example, are likely to be common across cells, and costs, capacity management, sourcing, and strategic direction need to be managed centrally. This lead to the development of a matrix organization similar to the one shown in Figure 7.4.

Our experience of mini businesses and matrix reporting structures is that they are successful in improving communications, customer service, motivation, and quality. They also reduce lead times and costs. Problems can occur if the matrix organization is too complex or if objectives in the mini businesses are not in line with those at the strategic level. Supply chain objectives at the senior level will include supplier performance issues and should be closely linked to what the cells are measuring and improving. If, for example, there is a consistent problem with the quality of a particular part used by one cell, the KPIs at both the cell and business level will be affected. The cell

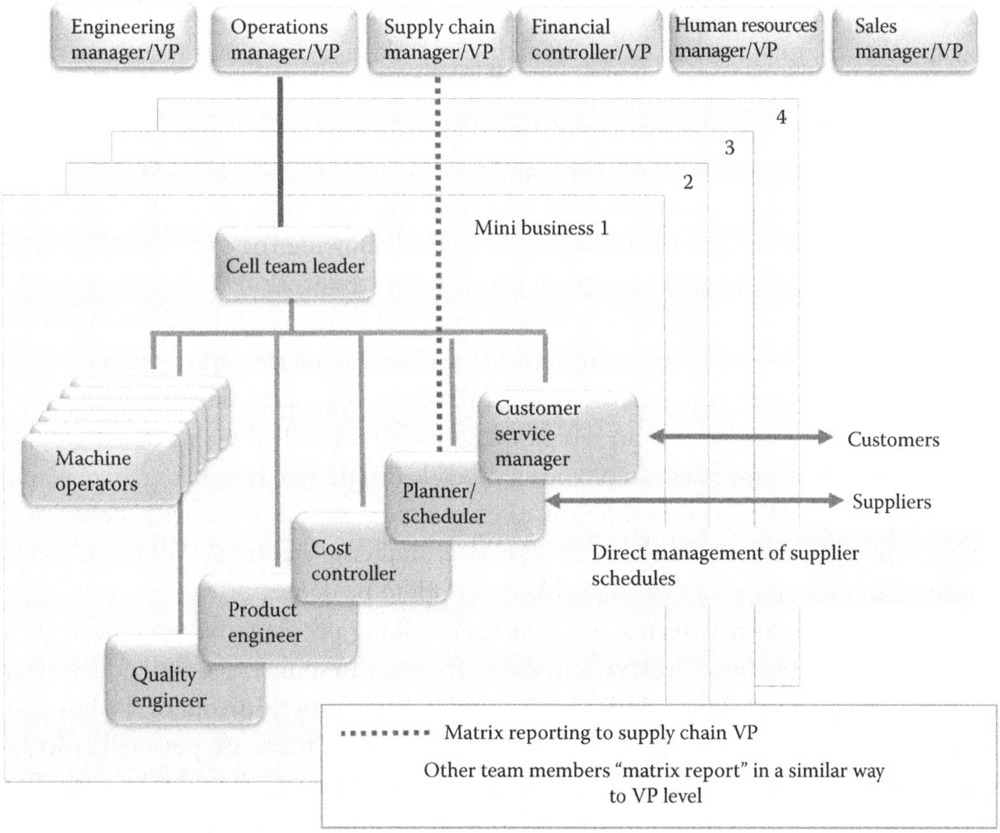

Figure 7.4 Reporting structure for cellular/mini-business organizations.

team will make an effort to try to fix the problem. However, they may require more skills or capacity from central resources in order to solve the problems. If the supplier is not responding well or is judged not to be in a position to make the necessary quality improvements, a decision to re-source the parts may be needed. These are examples of how the matrix organization has to work. If there is insufficient communication between cells and central resources or KPIs are inconsistent, the organization will not gain advantages from the new organization.

Cells or mini businesses can deliver the same improvements in communication and teamworking in a range of businesses. The most common application we know of is in manufacturing. However, here is an example of mini businesses operating in the aircraft maintenance sector.

A maintenance and repair operation for a supplier of aircraft assemblies supports units from a range of aircraft and airline customers. When units are received from customers they are stripped and examined and the work required is planned. This may involve making or buying parts, assembly, testing, and certification. The work required varies with each unit depending on its condition. The operation has always had a traditional departmental structure and there are sales, engineering, manufacturing, purchasing, planning, finance, and human resources departments. The flow of work and speed of communication across departments is slow, and as a result, customer lead times are long and unreliable. Customers also find it difficult to get the exact status of specific units. The business reorganized around maintenance cells with small central functions. A large proportion of the manufacturing, assembly, and test became dedicated to types of unit. In addition, direct contact with the customer for unit scheduling, feedback, and queries was moved to the cells. Following the initial inspection of units, the cell teams obtained the parts and planned the work required. As a result, supply chain communication was faster and more accurate. Lead times were reduced and customers were able to get more accurate and timely progress information.

Many other organizations have traditionally been organized departmentally and some are now finding benefits in reorganizing into multiskilled teams based around the work. Business processes can provide a valuable insight into how to organize teams around the work. When normal processes cross departmental boundaries, delays and changes in priorities are likely. This should prompt the question, Why not move resources into a team structure to enable this work to be done?

Reorganizing into new teams and sometimes asking people to be more flexible in the work they do can prove worrying to those involved. In some cases, individuals who have been doing a very focused, single task find it difficult to think out of the box and take the initiative to do something. People often suspect there are personal risks in changing the organization and need to be persuaded that they should take part. The persuading and preparation of people to join new teams and try the associated new ways of working is an important task. Some guidance on how to approach this are contained in Chapter 9.

In order to work effectively, teams need to contain a mix of different types of people. Some are inventive, some are good communicators, some have good attention to detail and some do not. It is important to make sure the team has the right range of people types; otherwise, it will not deliver results.

In order to get the team members to work effectively, I have found it useful to think about the Belbin Team Roles (created by Meredith Belbin, visiting professor at Henley Management College, Oxfordshire, UK). I recommend you go to http://www.belbin.com, but just to whet your appetite, see Figure 7.5, which summarizes his team roles.

This may seem a little overly complex and I'm not suggesting that you structure your entire recruitment efforts around Belbin. However, I have certainly noticed that teams that do not work, which on examination, are missing some *Belbin roles.*

7.3 Commodity Groups and the Organization

Commodity groups have proven to be a good organizational way to create teamwork with other disciplines around specific types of purchase (e.g., electronics, chemicals, moldings etc). Figure 7.6 shows part of an organizational structure that I belonged to once when I had a proper (non-consulting) job.

BELBIN®

Team Role Summary Descriptions

Team role	Contribution	Allowable weaknesses
Plant	Creative, imaginative, free-thinking. Generates ideas and solves difficult problems.	Ignores incidentals. Too preoccupied to communicate effectively.
Resource investigator	Outgoing, enthusiastic, communicative. Explores opportunities and develops contacts.	Over-optimistic. Loses interest once initial enthusiasm has passed.
Co-ordinator	Mature, confident, identifies talent. Clarifies goals. Delegates effectively.	Can be seen as manipulative. Offloads own share of the work.
Shaper	Challenging, dynamic, thrives on pressure. Has the drive and courage to overcome obstacles.	Prone to provocation. Offends people's feelings.
Monitor evaluator	Sober, strategic and discerning. Sees all options and judges accurately.	Lacks drive and ability to inspire others. Can be overly critical.
Teamworker	Cooperative, perceptive and diplomatic. Listens and averts friction.	Indecisive in crunch situations. Avoids confrontation.
Implementer	Practical, reliable, efficient. Turns ideas into actions and organizes work that needs to be done.	Somewhat inflexible. Slow to respond to new possibilities.
Completer finisher	Painstaking, conscientious, anxious. Searches out errors. Polishes and perfects.	Inclined to worry unduly. Reluctant to delegate.
Specialist	Single-minded, self-starting, dedicated. Provides knowledge and skills in rare supply.	Contributes only on a narrow front. Dwells on technicalities.

Figure 7.5 Belbin roles. (Reproduced from © BELBIN, UK, available at http://www.belbin.com. With permission.)

In this commodity group structure, each commodity group manager is responsible for

- Developing the strategy for his or her commodity (e.g., electronic components, electronics assembly, machined parts, plastics, software, services, and indirect purchases)
- Managing the key relationships
- Managing the buyers and sourcing people in the commodity group

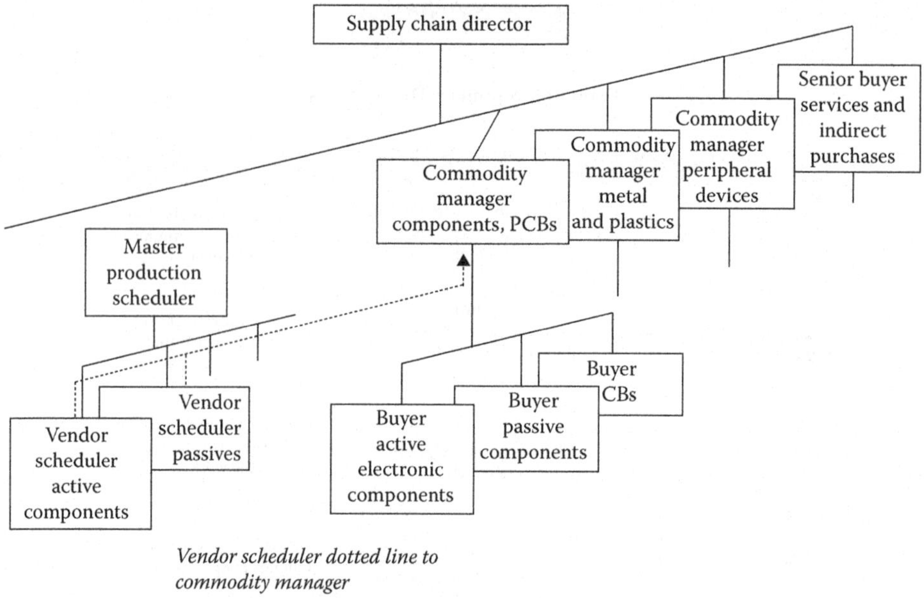

Figure 7.6 Commodity group organizational structure.

- Managing the contribution of the commodity group to the business in terms of cost, quality, and delivery improvement, and also their other contributions to the business

Commodity groups are teams focused on their own suppliers and technologies, but they also belong to temporary teams including manufacturing and engineering design team members in order to address specific areas or even tasks. A modern example could be the development of a process to manage the purchase of chip wafers, slicing and dicing, chip testing, chip storage, and chip mounting on modules. This type of process involves issues of supply chain, quality, manufacture, engineering, and a great deal of money and risk. Similar teams were often formed to agree on total costs, and select suppliers for application-specific integrated circuit (ASIC) chips and design tools.

Figure 7.6 also shows the dotted line responsibility of each vendor scheduler to the relevant commodity group manager. The vendor scheduler's primary reporting should be to a senior planning role. In my experience, this is often to a *sales and operations planning manager or master scheduler.*

7.4 Linking the Planning and Purchasing Functions

The *S&OP manager* or *master production scheduler* roles tie together the commercial side of managing the supply base with the management of schedules to meet customer needs. A lot of businesses we have worked with have implemented cellular organizations. In these cases, the tactical scheduling responsibility is usually transferred into the cellular teams similar to that in Figure 7.7.

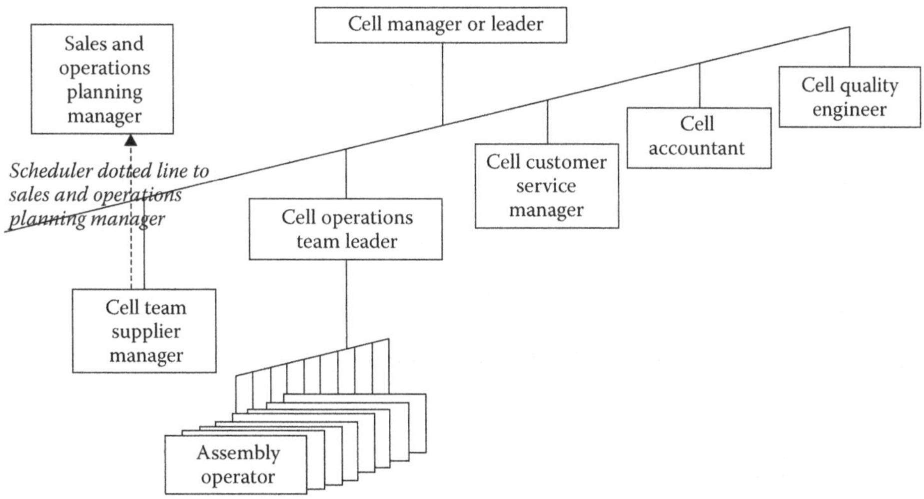

Figure 7.7 Cell or mini-business organizational structure.

These are matrix organizations and they can work very well for the supply chain, although conflicting objectives often have to be sorted out. There are many other parts of the organization in which supply chain people need to form teams and in my experience these teams often have to be formed, adjusted, and re-formed as business is done—for example, new product development (NPD). In reality, NPD projects often need a varying range of supply chain resources through the project—for example strategic sourcing, prototype procurement, supplier selection and on-boarding. The matrix organization is usually not flexible enough to cope with the rate of change in these cases and therefore other ways of managing transitory teams need to be used. The demand for supply chain people to do work for and belong to more teams and projects soon outstretches resources available and some combination of centralized and team devolved resource is needed. For example, a supply chain member of an NPD team might do some of the supply chain work, but also act as a coordinator of work being done by other people—e.g., Commodity Managers.

So far the supply chain can be seen to have involvement in a wide range of business issues. In many cases, these involve joint working across departments. However, in terms of basic supply chain responsibilities, here is a top ten list of supply chain questions for you to consider:

1. What do our internal/external customers need from us and how well can we provide it?
2. How competitive are we on prices and costs?
3. How competitive will our plans for prices and costs make us in the future?
4. How do we manage the balance of supply and demand?
5. How good is our inventory management at balancing customer service and cash flow?
6. How can we improve both the customer service and distribution costs?

7. How well do the supply chain processes work? (from forecasts and customer orders to delivery and invoicing)
8. How do key suppliers contribute to areas such as new technology, cost reduction, and performance improvement?
9. What do we contribute to the success of our suppliers and are they able to develop themselves to fulfill our future needs?
10. Do our supplier technology roadmaps provide what our business needs?

These questions will never be perfectly answered, but they represent constant goals for the supply chain organization.

7.5 Managing Up, Managing Down, and Managing along the Supply Chain

The supply chain survives on communication. If you are responsible for managing part of it, you are accountable for communication on tactical and senior-level issues. This often extends along the supply chain, as the saying goes, from supplier's supplier to customer's customer. Senior supply chain managers often spend a large percentage of their time talking to their people, their stakeholders, and suppliers making sure they have a good understanding of the situation themselves and that others also understand what they need to do. For example, if you were a supply chain manager:

- Would you want to know if one of your key suppliers had lost some business and was in financial trouble?
- Would you want to know if a design team had been discussing possible production options and volumes with a supplier without telling your supply chain team?
- Would you want to know if your manufacturing team was losing confidence in one of your supplier's ability to maintain quality levels?
- Would you want to know if your manufacturing team was losing confidence in the ability of your supply chain team to manage critical supply issues?
- Would you want to communicate directly to supply chain people about a radical new strategy that required re-sourcing away from many local suppliers to new ones in Asia?

The answer to all of these is yes, of course, but it takes a lot of time to be fully conversant with all issues. One approach that seems to work even if you are very senior in a very large organization is to develop a set of people whom you will be able to trust to tell it like it is. It is not a bad idea even if you are not that senior. Use your own judgment about the people you chose, but they must have proven to have good judgment themselves, be well connected, and are not afraid to tell you things you will not like to hear.

Flexible team-based structures help supply chain people communicate with internal functions.

* * *

8

Lean Supply Chain Practice

8.1 Applicability of Lean Tools in the Supply Chain

Juran and Deming, two American statisticians, helped Japanese industry to recover after the war. A fundamental part of their approach was based on the measurement of failures, the diagnosis of the cause of failure, and the implementation of remedies. The repeated application of this technique became the continuous improvement movement and the epidemic of quality that revolutionized Japanese manufacturing. The level of failures in Japanese electronics products had been reduced to unprecedented levels by the 1970s.

Toyota was relentless in its continuous drive to remove "waste" from its manufacturing processes. Its level of efficiency and quality also started to outstrip that of European- and American-based car manufacturers. *The Machine That Changed the World*, written by James P. Womack, Daniel T. Jones, and Daniel Roos, sold over 600,000 copies. The book did a great deal to promote *Lean* manufacture that was based on the Toyota approach to removing waste. The Lean approach has become a massive global movement. People from a wide range of industries have now adopted the sort of improvement practices used in Japanese car manufacturing.

Six Sigma is a term first coined by a Motorola engineer named Bill Smith (Six Sigma is a federally registered trademark of Motorola). Six Sigma is rooted in the statistical approach to achieving extremely low levels of failure (Six Sigma or 3.4 failures per 1 million occurrences). However, many of the Lean tools are used to achieve the reduction of failures.

Many organizations are using a mixture of these tools to achieve improvements. Some even use the term *Lean—Six Sigma*. Most Lean tools are as applicable to supply chain activities as they are in manufacturing operations. Three tools are summarized in Figure 8.1. They are perhaps the most useful for supply chain improvement. Process mapping (or similarly, *value stream mapping*) always delivers something surprising. Usually it is a part of a process we did not know about, an additional step or holdup. These constantly provide potential for supply chain improvement. The *wastes*, seven or eight of them, provide a vehicle for teams to *see* waste, define it, and remove it. The Six Sigma (DMAIC)* tool is similar to the Lean tool (PDCA)† and sometimes referred to as the *Deming cycle*. These are all forms of the fundamental process of continuous improvement.

---

* DMAIC, defined by Bill Smith of Motorola (Define, Measure, Analyze, Improve, Control).

† PDCA, made popular by Dr. W. Edwards Deming (Plan, Do, Check, Act).

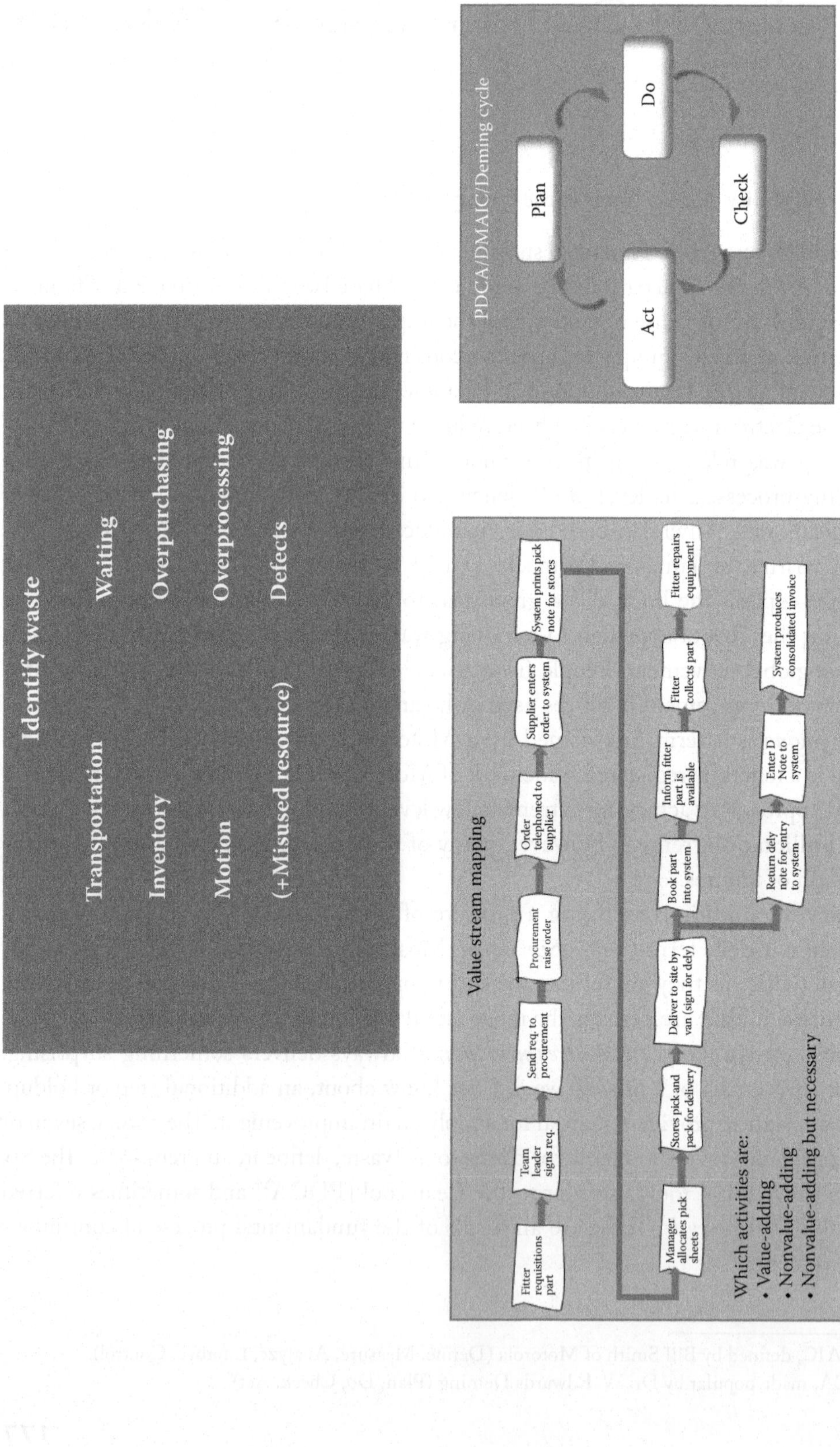

Figure 8.1 Lean and Six Sigma tools.

Lean tools can make a fundamental improvement to performance and are as applicable in the supply chain as in manufacturing.

* * *

The following examples describe some of the simple tools we have found most useful for supply chain improvements.

Performance is driven from the top. Direction, policies, frameworks, and outline procedures need to be defined with a strategic view. The delivery of excellence, however, comes from what we call *the way the work is done*. Teams who do the day-to-day work need to make detailed processes perform. Top-level management cannot make this happen. They can only create the right ingredients for the business's teams to do it. The Lean approach is one of the best ways we know to drive real improvement at the team level—and the supply chain is rich with opportunity.

8.2 Waste Removal

The waste removal concept can be simply put as the removal (or at least minimizing) of any activity that does not add value from the end customer's point of view. People now use seven or eight wastes, which can be remembered by a range of acronyms. The one I tend to use is TIM WOOD, as shown in Figure 8.2. The purpose of the waste

Transportation	Moving something from A to B does not increase its value from the customer viewpoint. Although (like many of the *wastes*) it may seem unavoidable, we can often find a local source, or ask a supplier to add a new location. Many of the decisions then come down to the total cost and response time. Minimizing the amount of transportation (mileage) in most distribution networks is a prime objective.
Inventory	Inventory normally exists because the lead time for obtaining supplies is longer than the customer expects to wait. Reduction of lead time is therefore an important way to remove inventory.
Motion	This relates to the movement of people—for example, when they need some standard stationery (e.g., paper), can they get it from somewhere (a cupboard?) close by, or do they need to walk to the next office to get it?
Waiting	There are many different examples of this waste. For example, a delivery vehicle is waiting for the unloading dock to become available, or parts are waiting to be assembled because we do not have all of the parts yet.
Overpurchasing	This is the supply chain version of making more than were required on the production order. There are many causes for overpurchasing—minimum order quantities, package sizes, quantity price breaks, container quantities, etc.
Overprocessing	This relates to paperwork or transactions that are not justifiable. Examples are too many multiple approvals, purchase orders for items whose value is less than the administrative costs of the purchase order, and use of manual transactions where information could be transferred using labels or tags or similar devices.
Defects	There could be a large list of examples here—besides the obvious delivery of faulty or wrong items, or wrong delivery address, etc. Process defects are good examples. Invoices that fail the three-way match are failures, inventory inaccuracy.
Plus the 8th waste usually described as misuse of resource	This is the misuse of people resource. Multiple approvals of purchase orders, manual raising of low-value purchase orders, manual transactions that could be automated, perhaps even manual labour that could be economically replaced by robots.

Figure 8.2 TIM WOOD.

definitions is to enable people to see waste. This is important because a lot of waste is often regarded as acceptable as part of the process. In reality, once you are familiar with any process or area of work, you are likely to fail to even recognize the wastes. Using TIM WOOD, Figure 8.2 explains each of the wastes in a supply chain context.

Some of these examples are repeated and can properly apply to more than one of the eight wastes. The important thing is to recognize the waste. Once you have recognized it, you are obliged to take action to remove or minimize it.

8.3 Process Mapping

Effective business processes are essential. Too often, they are overly bureaucratic, take too long, duplicate work, carry out work that is not needed, and do not produce the right result. Many business processes have grown and developed over a period of time and have lost their focus along the way. Processes that are driven by e-mails are quite common and unless these are systemized, the process lacks uniformity, is difficult to follow up, and can achieve variable results.

Process mapping is a valuable tool in the supply chain because our processes tend to interface with more areas inside the business as well as with suppliers. Clarity of process is essential; otherwise, the financial consequences are likely to be large. Culturally, even the process of getting people together to map joint processes can be beneficial—for example, supply chain and accounts payable people process mapping the invoice clearance process will almost certainly clarify some areas of misunderstanding.

If you are a manager, good processes should normally be part of your responsibility. When arriving in a new job, it should not be too long before you take a serious look at the business processes. If you find they are not clearly defined, you should start off some process mapping activity. If you ask senior management what the process is, they will tell you what they think it should be. If you look at the International Organization for Standardization (ISO) definition in the businesses ISO standards, it may lack detail and be out of date. The best way to do process mapping is manually with the people who actually run the process. I still favor a large wall and some sticky notes. This is fairly flexible; if the people doing the mapping remember something or want to change it, you can move the existing notes around.

One of the most useful ways to start process mapping is with a group of people who know the process, gathered around a large sheet of paper or whiteboard to construct the diagram. It is useful to collect the key documents or computer transaction screens before starting. Together with data on volumes and any relevant faults or KPIs, these can help to ensure the current process is captured and understood. Discussions with the team members who carry out each activity along the process are required in some depth: Is the input data accurate and timely, and does it meet the process requirements? Does the process activity achieve the right results effectively and in the right timescale? Is the output provided from each activity accurate and timely, and does it meet the requirements of the next activity in the process?

The difference between business processes, activities, and tasks can be useful to bear in mind. Complete business processes achieve an end result for the business, whereas activities and tasks are subsets or step along the process.

The map of the existing process is likely to unearth a number of potential improvements. Some of these will be easy to implement and deliver significant improvements. In addition, it is often useful when going on to map out the new or "to be" process, to have some unconstrained input—blue sky/out of the box/radical or whatever you wish to call it. We should not discount the ability of people who do the work to think like this, but it may also be useful to have some "fresh thinking" from people who are not normally involved. These could be from IT, other departments, management, or even customers. In some cases, the need for a process, or certainly parts of it can be in question. More likely, there will be new approaches, IT tools etc., which could radically change how the work is done. It is important to map the new process in some level of detail in order to test the idea, capture it and communicate it to others.

Process maps are also called value stream maps (VSM). When using this term, some people mean exactly the same as a process map. However, used properly, the VSM shows the adding of value and cost through the activities of the process. Waste can be identified and the relationship between cost and value can indicate where activities could be modified to bring them in line with the value they deliver.

Figure 8.3 shows a process for obtaining a low-value component for use on a prototype product under development. The process is pretty efficient. It uses a computer-based system for raising a requisition and getting it approved. Then, purchasing raises a PO directly on the supplier's website. The part is picked, packed, and dispatched the same day. A courier makes the delivery the following day and goods inward deliver the part to engineering. In Lean terms, the only *value-adding* activity then takes place—the component is used on the new product. The elapsed days are shown on the process map. For new product development, time is often the critical issue (not cost). Since the process is efficient already, there is little scope for improvement. However, if the

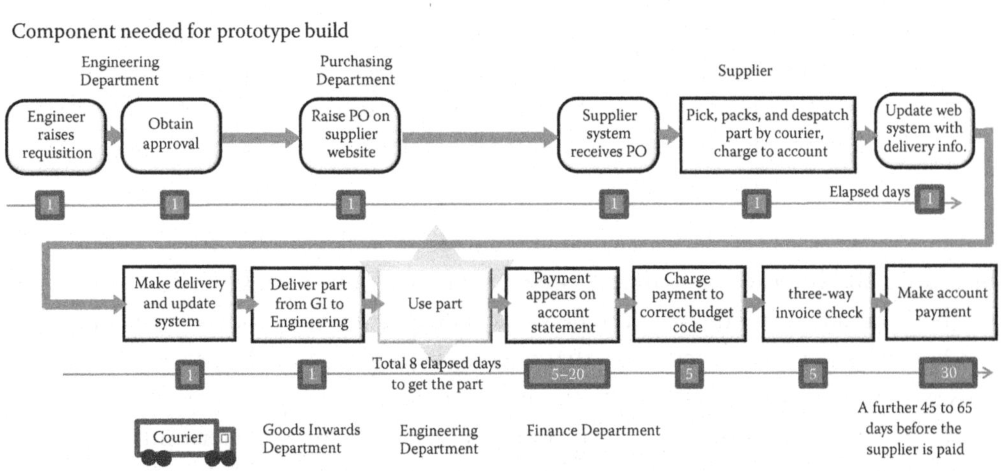

Figure 8.3 Process or VSM.

engineer were to be given a P-card, he or she could raise the PO directly on the supplier and 1–2 days of waste (50% of the total) could be removed from the process. Although the supplier will have to pay a charge to the Card provider, he will get his cash quickly. This is a good example of a radical new process which would bring significant benefits for low-cost items. It is surprising how the use of simple diagrams can help people to talk about processes and uncover and discuss areas where improvements can be made.

Another point about this process map is that it does not end with the part being used. It goes on to include the payment to the supplier. It is important to follow the process beyond departmental boundaries to its conclusion. The engineer may not be interested in making sure the supplier is paid—until he needs another part from him!

Almost all supply chain processes are worth process mapping. Below are some that I find are the most advantageous.

- *Customer order receipt → delivery → payment receipt.* This might start with an IT facility that the customer will use to select the product they want. There may be product sizes and options to select. Product availability will usually need to be checked and this may be a simple matter of checking and reserving free inventory, or it could be a facility to promise a shipment date. Many processes will collect customer payment at this stage, or payment could be invoiced after delivery. If the product is in stock, the process will continue through picking, packing, and dispatch. The delivery part of the process will then take the part to the customer. The customer will need to be kept informed at various stages in the process. Delivery must be formally recorded and a customer signature may be required. If payment has not been collected, invoicing and payment processing must begin.
- *Planning → inventory replenishment → supplier payment.* There are likely to be several processes here, depending on the type, cost, and lead time of the parts. They might start with MRP output, the approval of the PO, order progressing, receipt, receipt and matching of invoice, and finally, supplier payment. Frequently used, low-value parts may use a VMI process. In this case, parts of the supplier's own processes will require mapping.
- *New supplier required → sourcing → on-boarding.* This process is often done poorly and quite informally. Best-in-class processes would be called something like supplier life cycle management and would also include discontinuing business with a supplier.

In each of these processes there will be process parameters, objectives, performance measures, and improvement areas. Some examples are given in Table 8.1.

The parameters are important to define what capacities the process must cope with.

Objectives are the service levels that need to be achieved by the process.

Key performance indicators are the improvement areas and targets that are currently to be worked on. The quality improvement process of determining the root cause of performance failures and applying remedies needs to be followed. (See leading and trailing indicators in Figure 8.7.)

Table 8.1 Process Parameters, Objectives, and Performance

PROCESS	CUSTOMER ORDERS RECEIPT TO WEBSITE	REPLENISHING INVENTORY	SOURCING NEW SUPPLIERS
Parameters	900 orders/day	11,500 parts managed	10 new supplier requests/month
Objectives	Response time < 1 sec	Stock outs < 5%	Reject < per month Source < 4 months
KPIs	Customer dropout during process < 2%	PO raised < 2 days	Reduce total number of suppliers to 600

8.4 Continuous Improvement Cycles: PDCA* and DMAIC†

The continuous improvement methodologies driven by Juran and Deming provided the basis for a fundamental shift in quality levels. Now referred to as PDCA, DMAIC, or the Deming cycle, the approaches use relentless measurement, diagnosis, remedy, and remeasurement. The term "zero tolerance" has been used to describe it. Due to this methodology, Japanese manufacturing achieved levels of quality that were clearly better that their Western competitors. Since then, their competitors have learned and applied similar techniques, or in some cases, they have gone out of business.

These continuous improvement techniques are highly usable tools for improving supply chain performance. They are particularly useful for business process improvement.

8.4.1 Basic Steps

Identify and define the problem or type of failure you are trying to eliminate.

Use techniques such as Fishbone analysis to define the potential causes of failure. The whole population of failures or a valid sample need to be assessed and root causes attributed to each failure in the sample.

Use of a spreadsheet is helpful to list failures and attribute them a root cause number or letter, and then graph the number of failures attributable to each root cause.

Using this approach, you will have *diagnosed* the root causes of the selected failure and evaluated how many failures are caused by each.

Select the highest priority root cause of failure (e.g., the most frequent, most costly, or serious) and develop potential remedies that will prevent this particular root cause of failure occurring again. Techniques such as restricted pull-down boxes on data entry screens or component shapes which cannot be fitted together in the wrong way can be used here to ensure that the solution is reliable (i.e., this cause of failure cannot happen again). Measure the continuing failure rate to assess if the solution has worked.

When the remedy to the highest priority cause of failure has been eliminated, it is time to proceed to the next cause of failure and repeat the process. The goal (as you will have heard) is zero defects.

---

* As referenced earlier to Dr. W. Edwards Deming.

† As referenced earlier to Bill Smith of Motorola.

8.4.2 Example

The number of credit notes issued by a company to its customers was causing customer dissatisfaction. Credit notes are clearly a sign that customers are receiving the wrong goods, or the price or discount are incorrect, or the product is damaged or defective. The high number of credit notes is a problem, but it is a symptom. The causes, some of which have just been mentioned, are many and various. This is typical of all sorts of failure in products and processes.

The company set out to reduce the number of credit notes using a continuous improvement approach. All the credit notes issued over the previous 6 months were collected and analyzed. Each failure was attributed to the root cause of the problem. The highest cause of failure was the charging of shipping costs that should have been included in the price of certain shipped products. However, using the Fishbone principle, the real root cause of this was determined to be unclear wording on the terms and conditions regarding shipping charges used when customer placed their orders. Making a change to the wording and publicizing the change to salespeople successfully reduced the number of credit notes due to this cause to nearly zero in the following quarter. Having measured this success, the next highest cause for issuing credit notes was examined and diagnosed.

A range of supply chain processes have failures that impact the business and can benefit from this continuous improvement approach. Here are some ideas. What are the root causes and remedies for some of the following?

- Supplier failure to deliver on time?
- Late payment of supplier invoices?
- Credit notes from suppliers?
- Excessive part inventory?
- Warehouse inventory count inaccuracies?
- Indirect purchase order failures (internal customer dissatisfaction)?
- Return to depot customer delivery failures?
- Supply chain staff attrition?

In the credit note example given at the beginning of this section, the work was completed by a team of people who were involved in doing the work, led by a manager with experience of continuous improvement. Management can lead but not do the work because they have insufficient detailed knowledge of the process. Improvements carried out by the process team themselves are also likely to be more culturally acceptable. The important role for management is to guide, sponsor, and steer improvements to enable them to be effective in addressing the issues most important to the business.

Supply chain work is part of numerous business processes that impact performance and competitiveness. Examples of a range of important areas of supply chain influence is shown in Figure 8.4. At any one time, there will be issues and opportunities for improvement in a great number of areas, and one of the problems is knowing where to start.

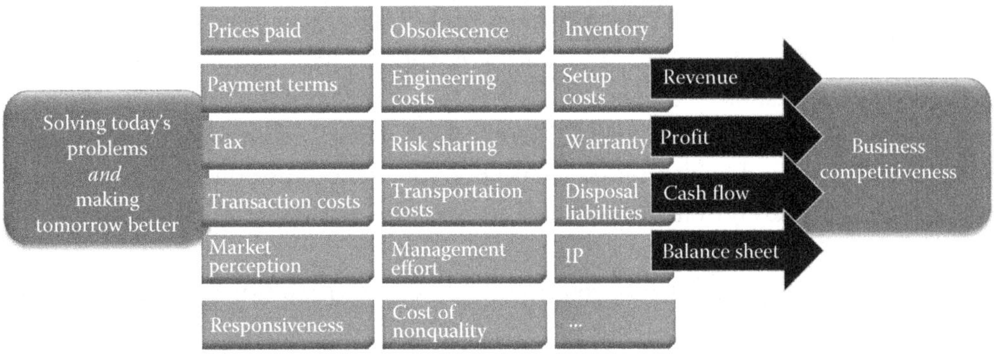

Figure 8.4 Areas of supply chain influence.

Supply chain professionals need to have a perspective of how different areas of performance impact the success of the business. This is essential to the value of the effort and work you and other team members put in. If you do not have this, your great work can turn out to have little beneficial effect on the business. This is not too hard once you have started to think about it. What could be the result of improved responsiveness in important supply chain areas—sourcing and placing orders for indirect purchases, agreeing on arrangements with suppliers enabling shorter effective lead times, faster ability to reschedule following changes in demand, and so forth—which of the potential improvements are both practical to achieve and will drive or enable the most improvement to the business performance? What could be the potential impact on Revenue, Profit, Cash Flow, and Balance Sheet?

Removal of waste, process mapping, and the continuous improvement cycle are very useful tools for supply chain improvement.

* * *

8.5 Example of Lean Improvements in the Supply Chain

You may find yourself in a new position, leading a supply chain team, perhaps similar to this example.

You report to the CEO and you have discussions with him or her about your new position and what the business needs you to deliver. The business designs and final-assembles a product for a worldwide market. Price competition is reasonably tough and work on supplier prices is continuous. However, the main issue on the CEO's agenda is that the supply chain organization is regarded as unresponsive and slow. The CEO quotes examples to you of a new product development program that required prototype parts urgently and yet the requisition waited 4 days in a buyer's in-box, similarly he quoted some urgently required parts had arrived in goods inwards and waited for several days before being "received." Several more examples are given before the CEO says that there are plenty of supply chain people and the company is looking to you to improve efficiency and responsiveness. Discussions with other senior people indicate the same type of issue.

You spend a little time in the key areas and observe that people are busy doing what appears to be good work. Since they are working hard and not achieving what is needed, you question if the work is being done efficiently. You decide to do some investigative work with some supply chain teams using Lean tools. You have used these tools before and you think they will give you a good introduction to your people and their work and enable you to find out why the department appears to be underperforming.

After some initial discussions, you decide to start separate teams to examine three areas:

1. Parts for new product development
2. Goods inward receipts

These were the areas in the examples given by your CEO, plus:

3. Rescheduling part deliveries following MRP runs

This is an area of complaint from the manufacturing team.

The teams have done some Lean work before and are familiar with wastes and with process mapping. These are the key tools used by the teams for their initial work, which takes 3 weeks.

The teams have produced maps of the main process steps using flip-chart pads and sticky notes and looked for examples of waste. The work has been done well and your involvement has been useful to provide you with an understanding of how things work. However, it is not yet clear what the key problems are and what can be done. There are a lot of examples of waste from the teams in each area. Some examples are shown in Figure 8.5.

After some discussions, some additional investigation is agreed on for all three areas. For the "parts for new product development" area, for example, "the whole process of requisitioning, purchase ordering, receiving, and invoice payment" is too general. A discussion indicates that the amount of paperwork is large and it is decided to quantify how many documents there are, for what type of parts, the quantity required, and the value. The results are surprising, especially when graphed as shown in Figure 8.6. The graph shows the value of purchase orders raised in descending order value for one NPD project.

Forty percent of the POs raised were for values of less than $100! All orders were subject to the same process. The administrative costs per order were estimated at $90 including receiving and invoice payment. They also represented approximately 40% of the administrative time spent on NPD orders.

You had seen P-cards working in a similar environment before and despite some initial opposition, it was decided to use them for NPD orders under $100. It was decided to make the NPD project manager the cardholder for all >$100 orders for their own project. The P-cards were set up with merchant codes to enable them to buy

Waste	Parts for new product development	Goods inward receipts	Rescheduling part deliveries following MRP runs
Transportation	Transport of items from overseas suppliers	–	Transport for multiple orders of the same part
Inventory	Engineers keep stock of parts "they know they will need"	Inventory waiting to be received	Some deliveries not cancelled in time
Motion	Supply chain representative walks to engineering to discuss requirements	Carrying goods received from dock area to receipts area	
Waiting	Requisitions waiting in in-tray	Inventory waiting to be received	Schedules waiting to be actioned
Overpurchasing	A full reel of 4000 capacitors was bought for making five hand-built boards		
Overprocessing	The whole process of requisitioning, purchase ordering, receiving, and invoice payment	Some low-priced items have been received five times during the month	Some parts are expedited one MRP run after they were de-scheduled
Defects	Three deliveries of the wrong parts—one was ordered incorrectly	Wrong receipt data entered—range of examples	Two examples of urgent orders due to planning parameters but not actually required
Misuse of resource	The NPD supply chain representative takes prototype parts to engineering	Parts relabeled due to damage on originals	Senior buyer helped to clear backlog

Figure 8.5 Examples of waste from three areas.

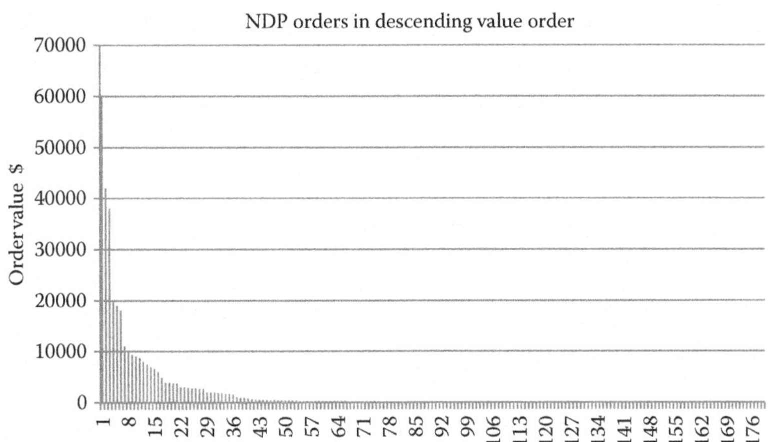

Figure 8.6 Profile of NPD parts for purchase ordering.

mechanical, electrical, and hydraulic components and with individual order values limited to a maximum of $100. Despite some initial problems over delivery points for goods, the P-cards were quickly seen to be a faster way to obtain parts and get them with a high level of security and accountability and with much lower administrative burden. The reduction in the number of POs in the NPD buyer's in-box also meant that the higher-value orders were processed more quickly, too.

Your work with all three teams continues to make improvements—not all as immediately obvious in their effect, but making clear the value of the Lean teams approach to improvements.

A range of improvements are subsequently driven by the three initial areas of Lean work, including

- Supply chain NPD representative colocated with NPD for 3 × one-half days per week, resulting in fewer journeys to NPD department, better involvement, and solving of supply chain issues
- NPD stock room setup, resulting in proper controls (how much of what have we got?) and storage (important for some delicate integrated circuits)
- Arrangements made with PCB manufacturer for supply of end-of-reel standard passive components (free of charge), resulting in immediate availability of standard passives
- Changes in MRP ordering parameters, resulting in fewer repeated reschedules
- KPIs used to measure receipt -> booking time
- Small nonpowered roller conveyer installed to move received boxes to receipts area, resulting in less walking and a well-organized sequence of goods awaiting receipt

These changes have not cost very much at all. Yet, together they have started to make a difference in measured and perceived performance of the supply chain. There are knock-on effects because people are spending time on more important tasks and a sense of teamwork and level of confidence that problems can be overcome has developed. If life is fair, this is also the time when the CEO tells you that you have made an impressive start!

Lean tools can be used internally or together with suppliers to deliver a large number of incremental benefits that can transform business performance.

* * *

8.6 KPIs

Here are some of the most common supply chain KPIs we know:

- *Delivery on time or on-time-in-full* (OTIF) is one of the obvious supplier delivery measures. However, in many businesses it would be considered disastrous to even have to measure it. OTIF would be expected; otherwise a production line or chemical process would come to a halt. OTIF is generally used when performance is bad and in a recovery situation.
- *Purchase price variance* (PPV) is an almost standard measure produced from the purchase ledger part of the finance system. It measures the latest price against previous prices for the same item. This provides a good guide to how

prices are increasing or decreasing and is useful if commodities purchased are relatively stable. Although this is useful, the cost may be significantly different from the price, in which case PPV can be said to give a misleading result.

- *Spend* divided by *revenue* or *spend* divided by *cost of sales*, while being crude measures, can be a relatively direct indication of the effect of pricing and costs on business profitability. However, it is too easy to come to the wrong conclusions if, for example, sales have been selling at higher prices or the cost of labor has increased. These are not therefore very usable as measures of supply chain performance, but may be useful at board level as an indication of the changing shape of the business model.
- *Days sales outstanding* (DSO) and *days purchases outstanding* (DPO) are measures of how many days' worth of invoices our customers owe us and how many we owe our suppliers. If you add consideration of the inventory level, you have a measure of how well the business uses its cash. Stock market analysts use the cash-to-cash cycle (DSO + Inventory – DPO, all expressed in days) partly as an indicator of how quick and effective a business's processes are. Financial people also like to balance DSO and DPO. Cynically, this means they try to delay supplier payments until they have the customer's cash. One French hypermarket chain is indeed famous for its fast collection of customer cash that enables it to fund expansion. In the final analysis, efforts to reduce all parts of the cash-to-cash cycle are good for the business.
- *Supply chain savings* are partly covered by PPV (above), but the savings can also include costs rather than just price (as in PPV). For example, if we arrange for a supplier to manage inventory for us (VMI) directly on the shop floor, we may reduce the costs of many inventory management transactions, but his or her price may stay the same or even go up. The total cost effect will be good, but PPV will be poor (the PPV impact should be low because these are low-cost parts). There are a lot of similar situations and a cost model may be needed to persuade people of their merit. When putting together the model be careful to include only savings you can clearly show to be real; otherwise, your credibility along with the supply chain will be damaged.
- *Number of suppliers.* It is a well-known assumption that we all have too many suppliers—and it is usually true. The *spend report* for almost all businesses produces a highly pronounced Pareto curve (more than 80% of the spend within the top 20% of suppliers). Too many suppliers for high spend items can mean diluted leverage. Too many suppliers at the low spend and long tail of the Pareto curve often means excessive transactional costs. While this can be an effective measure, be sure it is fully understood (see the four-box strategy [Kraljic-type analysis] in Chapter 4).
- *Purchase orders raised per day* are seen by some as a measure of how hard/efficiently their supply chain people are working. It can indicate a view that

effective supply chains are about efficient processing of administrative tasks. This is only true to a limited extent. Fortunately, supply chain people are smart enough to use every opportunity to increase their order throughput without working any harder. This KPI is of the old time-and-motion-study era and is not recommended.

- *Customer satisfaction* is difficult to measure without resorting to surveys that tend to be subjective. However, they are still valuable. Surveys that gather more detailed information without being too lengthy are most useful. These can at least ask about satisfaction in areas such as response time, communication, and feedback. Some space for free format comments is often even more useful. Also, what about a survey of supplier views—once the initial glossy compliments have been dispensed with, there can be some useful comments and "home truths."
- *Lead time* from order receipt to delivery is critical to supply chain performance. However, it is difficult to measure and can be variable depending on supplier workload and capacity. Our own customers usually expect a dependable and fixed lead time.

These measures are largely to monitor whether performance is acceptable. The following measures are part of a performance improvement effort. They will be instrumental in helping to improve quality. After the improvement has been securely achieved, these measures will be dropped.

- *Number of POs raised under $200.* These were a major driver of late invoice payment. The problem was largely solved by using P-cards and VMI. Following successful implementation, the measure was dispensed with.
- *Number of purchase orders placed within lead time.* These were a major cause of poor delivery-on-time.
- *Number of unplanned stock adjustments made.* These were found to be frequently done too hurriedly but not actually required. They were a major cause of inventory inaccuracy.

One of the most important purposes of measurement is to drive performance improvement. When I worked for Research Machines, the cofounder, Mike Fisher, believed strongly in Juran's* approach to quality improvement (this was some time before the famous Lean approach and the book *The Machine That Changed the World* by James Womack, Daniel Jones, and Daniel Roos). We were trained in the Juran approach, working in teams, learning to do quality improvement projects. Since then, the importance of measurement as a diagnosis and improvement tool (the diagnostic and remedial journeys) has always been with me. It has enabled me to get to the root

* See opening Section 8.1 regarding Joseph M. Juran and Dr. W. Edwards Deming.

cause of issues, fix them, and measure the consequences. When I worked for Ingersoll Engineers they had a similar approach:

- Measures must motivate ACTION
- Actions must bring MEASUREABLE RESULTS
- Results must be TARGETTED

This is also similar to the Lean's PDCA and Six Sigma DMAIC tools. However, performance measurements are used for a range of purposes:

- The best use of measures is as described—to promote and enable an improvement program of some sort. An example could be a team working on reducing the number of incorrect inventory cycle counts each week. At the outset, 4% of the parts counted could be wrong, but the target might be <1%. After the team has completed their first piece of work this could be down to 3.5%. If the measures continue to go reasonably quickly in the right direction the work will be a success. Background on how the improvements have been achieved will add further confidence in the work and people are likely to support the program (see discussion on leading and trailing indicators later in this section and in Figure 8.7).
- Sometimes measures are just to monitor—to check that a process is in control. This is similar to a control chart on a machine tool. The process is working acceptably while the measure is within limits. Measures outside limits indicate that the process needs to be stopped and corrected. Some supply chain examples could be
  - How many action messages are produced from successive MRP runs. If a large and unexpected change occurs, it may indicate that something is going wrong and call for investigation.
  - Effective lead time (measure receipt date—PO placement date). This could indicate if supplier capacity is becoming more/less available.
  - PPV. This indicates market trends and purchasing success in price reduction. It is a good indicator because it can help to foresee changes in cost of sales and profitability.
- Performance measures are also often used to promote a cause (e.g., the department has been successful in meeting its target again this month). This can be useful when there are tough new targets to meet, but if the same sort of target is met every month, it becomes repetitive and it may be time to set more demanding targets.

  We need a range of measures for different purposes. They are important because they will drive performance improvement and form a dialog with internal and external customers about our performance.
- Internal and external customers have a range of needs, often referred to as critical to quality (CTQ). These may conflict or change over time and are often not quite as we expected. They are, however, important to understand

and meet. Do customer surveys help? They can give a general view of customer satisfaction, but if the results are poor and no further insight is given, they can be frustrating. Some surveys ask for more details, but if we ask the wrong questions, the replies will also be unhelpful. Surveys that enable at least some of the answers to be in free format can give us insights and information that we had not considered. However, we are likely to need to spend more time analyzing them—recognizing common themes and grouping comments together. Surveys can be the start of a listening process with key customers or suppliers. We need to follow up key themes we do not understand and find out the real underlying issues that need to be measured and improved. Our supply chain business objectives are usually around total cost, quality, and customer service levels. However, it is important to recognize that priorities vary substantially in different business areas. NPD programs may need to get components and materials very quickly to support the project's *time to market* objectives. A simple measure of this would be the time from a request to delivery (to the project, not goods inward). Large-scale projects might be better served by time-definite deliveries to project plan rather than short lead time. In this case delivery-on-time may therefore be a better measure. In reality, they will often require materials to be available to plan, but delivered when called for on-site. If you identify the needs of key customers, you will probably come up with a range of CTQs and metrics. They will relate to end customer needs, possibly about spare parts supply as well as normal customer service. Sales may need cost, quality, and lead time, but also perhaps something more innovative. For example, faster availability of new products, more product options, or innovative new delivery arrangements. There will be general business objectives related to strategy and improvement targets. Suppliers may also have some needs, the most obvious being payment on time, but different supply schedules could also be on their list. Meeting these may also form part of the annual supplier negotiations.

KPIs are central to driving performance improvements. KPI could easily stand for key performance improvement.

Even the act of measuring and displaying the performance data for an area will encourage improvement because it is immediately perceived to be important. Most individuals have pride in seeing improvements and will react positively. They should be chosen carefully. Using too many KPIs will reduce focus on the most important improvement areas. A maximum of about six KPIs can be used for a particular business area in order to have this effect of improvement focus. More than this and there is a danger of overcomplication and dilution of the effect.

As an example, a business was very poor at paying suppliers on-time. The data and KPIs showed this every month; creditor days were constantly too long and the percentage of invoices paid on time was always unacceptably too low. However, these

KPIs were unable to drive improvement and a senior manager analyzed the potential causes of the problem. The results of his analysis are shown in Figure 9.4, in Section 9.4.2 about the use of two-axes graphs. This indicated that only 3% of the value represented 80% of the invoice volume. With the inevitable issues of mismatches and discrepancies that need to be managed when clearing invoices, the accounts payable team was swamped with work that delivered little value. The senior manager realized that the problem could only be tackled by reducing the number of low-value invoices. This could be done, for example, by rationalizing suppliers, making arrangements for consolidated monthly invoices, and implementing VMI processes for low-value items. The graph shown in the two-axes analysis is not a KPI, this is the analysis. The KPI used by the manager was number of invoices under $500. This was applied to supply chain managers with reducing targets. The result was that the reduction in low-value invoices started to enable improvements in supplier payment on time.

This is also a good example of leading and trailing indicators, as shown in Figure 8.7. The aspect of performance we need to improve is often a symptom (trailing indicators). It is important to establish the root causes, find solutions, and then measure the results (leading indicators). This will result in the improvement we seek.

In the example above, the objective was to reduce creditor days (trailing indicator). The most significant key to achieving the improvement was reducing the number of low-value invoices (leading indicator), and consequently, the massive amount of low-value-adding time they created. Once the objective improvement has been achieved, it is important to stop measuring the KPIs. They will proliferate and having too many KPIs diminishes their impact—they are no longer key. It is therefore important to manage the number of KPIs in use, and as mentioned previously, more than six risks a loss of focus.

Care needs to be taken when setting up KPIs because they can sometimes result in unforeseen and unwanted consequences. Our Figure 8.8 is lighthearted, but unfortunately, we have encountered some examples which are remarkably similar. Some examples we have encountered are

- Measuring the number of POs raised as an indicator of how hard buyers are working. This can result in an increase of smaller orders raised, which is absolutely not what we were seeking to achieve.

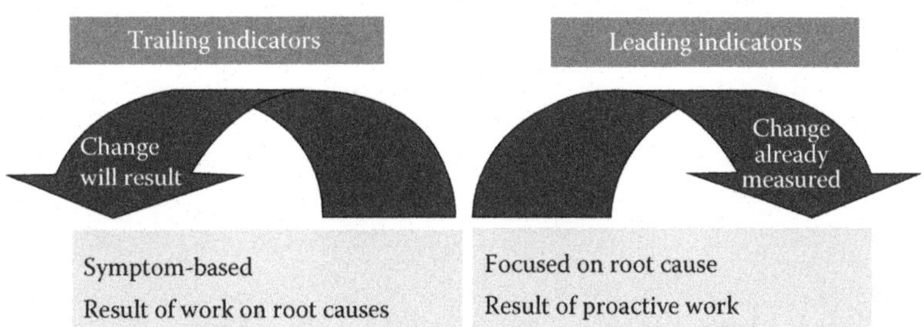

Figure 8.7 Leading and trailing indicators.

Figure 8.8 Bad KPIs drive performance in the wrong direction.

- Measuring the lateness of a product as it is dispatched from a factory. This led to some orders never being shipped because they were late, but would only be counted in the KPI when they were shipped. These orders were often canceled.
- Targeting price savings on parts and ignoring costs. (This can encourage sourcing to use suppliers from whom the quality or lead-time may increase total cost.)

Be careful that your KPIs do not encourage the wrong behavior.

* * *

9
Making Change Happen

When we want to change the way we do something, we need some level of support from the people who manage and from those who actually do the work. This may mean very senior levels of approval for large and costly change, but it is the people who do the work who will be directly affected. To a significant degree, they will be the ones who make it a success.

Some people are most helpful when changes are being made. Some people can be very obstructive. The difference between working in a team that pulls together rather than in different directions is immense on a personal level, but can make the difference between success and failure. Knowing how to persuade people to accept your case, and even more, to actually help positively with making the change, can often be more useful even than getting the solution perfect.

These mostly well-known approaches have been very useful to me and to my colleagues—helping us to persuade people to accept and become part of a change to what they do.

9.1 Dealing with Unpleasant News

With improvement and change comes unfortunate news for some. Principally this is loss of employment or changes in working conditions or position. There is often no way of avoiding this and I believe it is best to be as open, honest, and as helpful to those involved as we can be. Some useful guidelines are

1. Check the legal situation and be sure that you are meeting at least the minimum conditions. The law and custom and practice are very different in, say, the United States, United Kingdom, or Germany. Laws and regulations generally apply to
   a. The notice of change that people need to be given. For example, in Germany, this is at a much earlier stage than in the United Kingdom or the United States.
   b. Compensation, which generally relates to time in employment, salary, and contract of employment.

2. Tell people collectively, and then where appropriate on an individual basis, about the reasons the changes are being made.
3. Tell people about the plan and time scales for the change.
4. Tell people what it will mean for them individually in terms of
    a. Their long-term employment position:
        i. Does their current position continue?
        ii. Are they to lose their position?
        iii. Can they apply for a new position?
        iv. How long will their current position continue?
        v. Are there special things that they can do in their current role to assist with the change?
    b. Their compensation—how much will it be and how was this sum reached?
    c. What assistance will be given to help them find a new job?
    d. What is the effect on other personal terms such as pension, medical, and insurance coverage?
5. Identify people who are important to achieving the change, but who may or may not be needed afterward. These people need to be secured, usually with some special compensation.
6. After the initial meeting(s) to inform people what is happening, it is important to start activities such as project work on the change and assistance to people who will need to find a new job.
7. In some cases, particularly in the United States, common practice is for people to leave immediately, clearing only their personal belongings from the office. Where this is the case, I believe some form of support and contact needs to be continued to ensure that people are coping as well as possible with the change.

9.2 Step Change and Incremental Change

The difference between step and incremental change used to be characterized by the difference between industry in the West and in Japan. Since Womak and Jones and the Lean movement, the difference is less pronounced, but it still exists. Figure 9.1 highlights the difference between the two.

The step change in business performance is likely to be a management-driven change, for example, an IT system, investment in new equipment, or the consolidation of factories. The incremental change is more likely to be the result of tactical improvements, probably driven at a lower level of seniority. This is a typical result from good Lean practice.

The results from the incremental change business are likely to be more sustainable and at lower risk. On the evidence of Figure 9.1, the incremental change business would seem to be the better business because its performance is better for more of the time. In reality, the need for both types of change are driven by your business strategy and the level of competition.

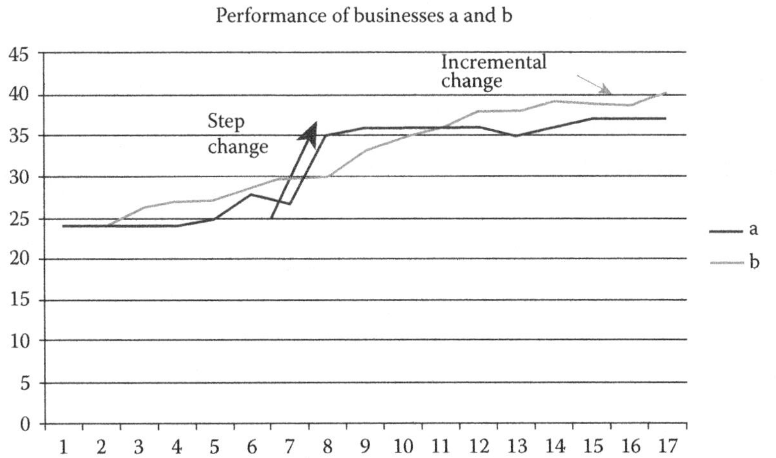

Figure 9.1 Step change and incremental change.

9.3 Taking Advantage of IT as a Driver of Change

New IT solutions are a frequent driver of change. They are usually management led and often involve large investments of resources and money. However, many of them fail to deliver fully the targeted benefits—nearly half, according to recent studies. The reason for this is not usually the technology, it is the people. Through wrong leadership or *cultural issues*, the new IT does not work well for them. We have seen a number of IT implementations where good concepts and IT functionality have been provided and the necessary training given. The implementation is then thought to be finished. There is a lack of connection between the IT project and the people who do the work. Either the IT program does not really understand what the new process should be, or they have not communicated it to the people who will do the work.

The focus on IT is sometimes therefore unhelpful. Focus on the new process is more likely to bring success, particularly with involvement of people in a similar way to Lean programs. For some reason, new processes lack the kind of enthusiasm and appeal that new IT does. So even if the title of the program sounds like IT, make it about the process and people.

Figure 9.2 contrasts examples of the significant IT issues with people and process issues for some well-known improvement programs.

The IT aspect of improvement programs can bring focus, enthusiasm, and gravitas from senior management. Getting the process and people issues in the program right can greatly increase its likelihood of success. If you bring these together well, you will revolutionize the way your business operates.

Program Title and Some Key IT Elements	Potential Key Process and People Issues
Sales and Operations Planning There are plenty of IT issues to solve such as • Integration between sales, forecasting, and planning data • Forecasting models that are more accurate	• Involvement of senior people in the S&OP process • Timely arrival of forecast data • Discipline to refrain from changing schedules within lead time
MRP • Implementation of such a broad range of applications • Large number of customizations to be managed	• Accuracy of BOM and inventory records • Getting VMI, kanbans to work with MRP • Getting users to understand how the system will respond to changes—in demand, planning parameters, inventory losses, and engineering changes
Warehouse management • Integration of WMS with automatic handling equipment	• Safety • Getting good processes going for cycle counting, etc. • Housekeeping? • Job losses if automating
Web-based sales for direct customer access • Interfacing with other systems for stock availability and to record sales • Generating data that provide understanding of how customers are using the site and how they can be attracted to *buy*	• Agreeing presentation *styles* with management • Getting customers to *buy*

Figure 9.2 IT-focused programs with key people and process issues.

9.4 Speak with Data

In the absence of facts and data, who wins the argument, or put politically correctly, whose views prevail? The answer is usually the person with the strongest personality, or shouts the loudest, or who has the most senior position. This is unlikely to guarantee that the right decision will be made.

However, being able to present facts in a way that clearly support your case can usually win the day. This is not always as easy as it sounds. There are two key things to remember:

1. There are likely to be those among your audience who do not support your idea. If you present data, they will try to discredit it. A well-known management technique is to pick away at the corners of the data until they find a point where the accuracy is questionable. When they can expose this, they question the whole of the data and your case is about to be lost. *Be prepared*:
   a. Know your data—where it came from, which databases, which time period, which elements of bad data were cleaned out, how accurate it is.
   b. Validate your data with other sources. If you are presenting to accountants, present or be able to show how the data agrees with their numbers. Show how your numbers correlate to authoritative sources, such as government figures, or surveys done by well-known bodies, such as industry associations.
   c. Ensure the relevance of your data—the customers/geographic areas/seasons/product types and so forth the same as in your case?

2. Your data analysis, particularly the way you present it, needs to make your case clear very quickly.
   a. Use the numbers. Some among your audience will naturally relate to them. They will provide an *anchor* for any other charts or graphs you use.
   b. Use graphs, tables, and charts that show or support the whole case in summary form if possible.

Your facts and data could be presented in a wide variety of ways, but there are some obvious graphs and techniques that are very popular—because they work. We cover my favorite three below: Pareto, twin-axes graphs (usually including Pareto), and paired comparisons (which provide a level of legitimacy in subjective areas).

9.4.1 Pareto Analysis

Pareto analysis, named after its Italian originator, is famously called the 80:20 rule. How did he work this out without the use of spreadsheets? The Pareto effect is seen in so many areas. It helps us identify the critical few from the multitude. It is the starting point for many very useful supply chain analyses:

1. A supplier spend report shows where the money is spent.
2. ABC classifications start with a descending value of usage report.
3. Inventory analysis shows the parts for which we have the most inventory.
4. Imposing analysis 3 onto analysis 2 indicates the relationship between usage and inventory values. This often shows that the lower usage or slow movers have relatively far more inventory than the fast movers—often called the sludge chart (Figure 9.3).

Supply chain data can be mined from the ERP system or elsewhere relatively easily. Some simple steps are

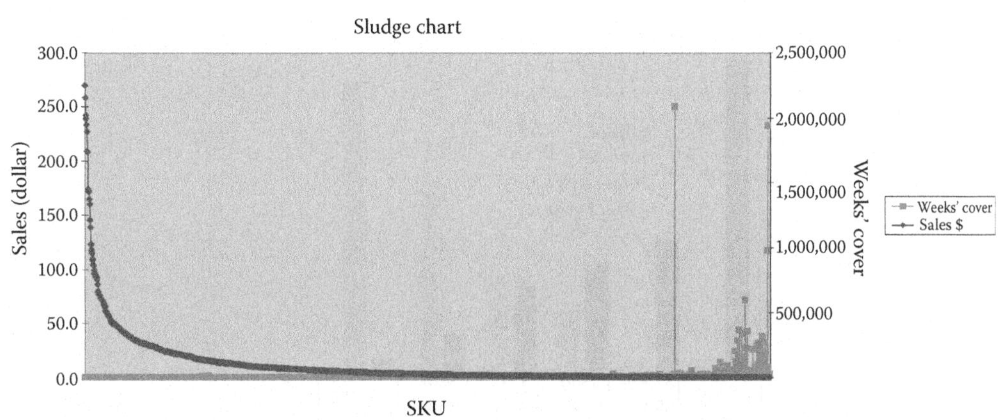

Figure 9.3 Weeks' inventory cover plotted against descending value of stock-keeping unit (SKU) annual sales.

- Be careful to understand exactly what the data is. For example:
  - Clean the data. A little while spent with a spreadsheet can pick up obvious errors—dates entered into the quantity field, unit of measure or currency differences, multiple entries, and so forth.
  - Check the data with another set of numbers you know to be right—is the total figure right?
  - Then see what the data tells you—speak with data.

The Pareto effect is simply produced by sorting in descending value of the field of interest—such as purchase order value, spend with each supplier, and so forth. The critical few are revealed. You can also add a column to calculate the cumulative value and percentage of the total on successive rows. Then you can see, for example, how many lines/records represent 20%, 50%, 80%, of the total.

9.4.2 Twin-Axes Graphs

When we have a basic Pareto showing, for example, that lower sales items have proportionally higher inventory levels, the contrasting of data on two axes can produce useful results, as shown in Figure 9.3.

This example, often called the sludge chart, shows the annual sales in millions of dollars for parts in descending value order—a Pareto shape. For the same SKUs the number of weeks' inventory cover is plotted against the right-hand scale. The sludge is the inventory on the right-hand side for which there are around or over 2,000,000 weeks of inventory cover (i.e., at normal usage, the current inventory will last 2 million weeks). The chart has been drawn to emphasize the point that there is a lot of slow-moving inventory for low sales items. After recognizing this, action can be agreed on.

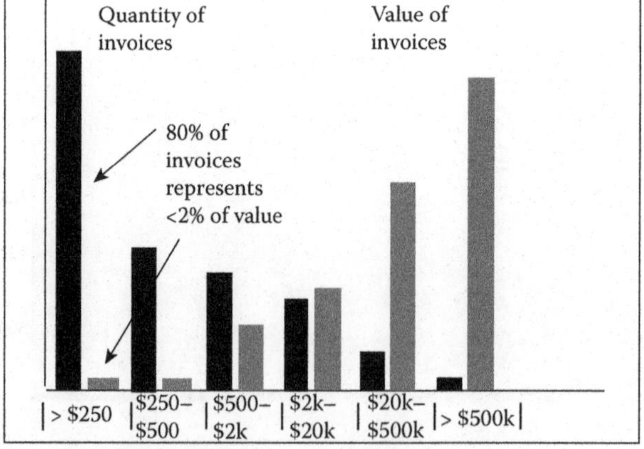

Figure 9.4 Value versus quantity of invoices.

Figure 9.4 tells us that most of the invoices we pay (80% of them) represent only 2% of the spend value! A similar story is likely to apply to POs, receipts, and warehouse picks. We spend most of our administrative time on items of insignificant value. This sort of analysis has been the starting point for many VMI implementations.

This is one of my favorite types of graph. There are so many useful contrasts of data that convey issues so graphically; they really do help you to speak with data.

Pareto analysis is one of the best starting points for analysis—it will nearly always lead to recognition of an issue. The addition of a second contrasting issue plotted on the other axis often tells a powerful story. Supply chain professionals can all use data analysis and presentation skills like this to persuade management of their arguments and get agreement on action to be taken.

9.4.3 Paired Comparisons

Paired comparisons are useful for determining the relative importance of factors that may be quite complex. It is also helpful to get people to agree on the relative importance. One of our most frequent uses of paired comparisons is for ranking supplier selection criteria.

The technique works by making individual comparisons of each pair of issues, one by one. The relative importance of each pair of issues is marked on a matrix that generates a *score* representing the overall priority of each of the issues.

For example a supplier evaluation was to be made for suppliers of machined and fabricated parts. Criteria for the assessment of quotations were grouped under six headings, each with a maximum score adding up to 100: max score (see Table 9.1).

The paired comparisons technique was used to agree on the weighting to be applied to a set of subcriteria for logistics capability for the site visit.

The criteria on each line were compared with each of the other criteria and the resulting score filled into the column. Score 1 if this criteria is more important than the one numbered in the column, score 0 if it is of the same importance, and –1 if it is less important. We added 10 to the totals to calculate the adjusted totals column, which makes all the scores positive numbers. The calculated weighting multiplies the adjusted total by 20/80 in order to achieve a total of 20 for logistics and the final weighting simplifies the weight to a whole number.

The results for the logistics heading are shown in Figure 9.5.

TABLE 9.1 Criteria and Weighting for Evaluation of Quotations

CRITERIA	WEIGHTING
Competitiveness (based on tender response)	20
Manufacturing capability	20
Quality	20
Logistics capability	20
Agreement and contract terms	10
Design assistance	10
Total	100

Logistics	1	2	3	4	5	6	7	8	Totals	Adjusted totals	Weighting	Final weighting
1. Organizational capability of personnel		0	0	1	−1	1	1	1	3	**13**	3.25	4
2. Level of process capability	0		0	1	−1	1	1	1	3	**13**	3.25	4
3. Planning capability	0	−1		1	−1	1	1	1	2	**12**	3	3
4. Inventory management capability	−1	−1	−1		−1	1	1	1	−1	**9**	2.25	2
5. Customer service level	1	1	1	1		1	1	1	7	**17**	4.25	5
6. Supply base size and focus	−1	−1	−1	−1	−1		1	1	−3	**7**	1.75	1
7. Supplier assessment and selection	−1	−1	−1	−1	−1	0		−1	−6	**4**	1	0
8. Supplier performance measures	−1	−1	−1	−1	−1	−1	1		−5	**5**	1.25	1
	−3	−4	−3	1	−7	4	7	5		**80**	20	20

Figure 9.5 Paired comparisons.

9.5 Change Cycle

This happens to us all. The curve shown in Figure 9.6 is adapted from work done in bereavement counseling. It shows how being confronted by change alters our behavior and performance as we adapt to the new reality. This applies to big changes, such as changing jobs, location, computer system, or organization, or to much smaller ones—even moving your office.

As someone presents new information or a proposal in a meeting you may have noticed this sort of reaction starting to happen to you. There is often surprise. This is often accompanied by a short period of shock during which you are unable to carry on as usual. The change dominates your thoughts; you may go to the coffee machine to discuss it with your colleagues or go somewhere quiet to take it all in. This can be short-lived because it can be followed by the well-known denial stage. Shortly

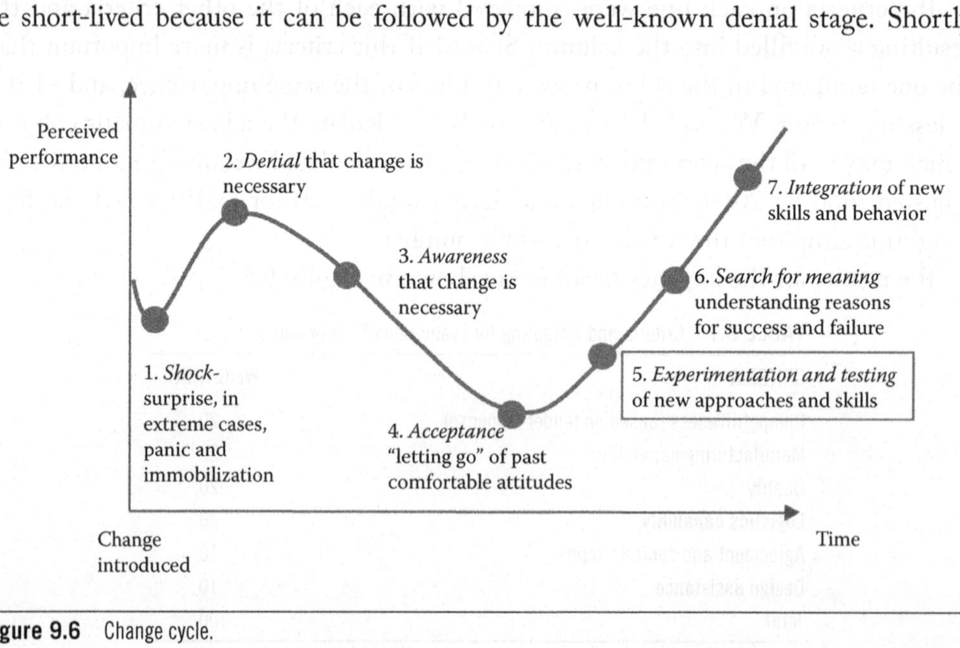

Figure 9.6 Change cycle.

afterward we think, "do not panic and carry on" when the change does not seem viable or at least will not apply to us. After that, there may be quite gradually increasing awareness of the reality and relevance of the change. Acceptance marks the lowest level of out perceived performance—how well we are operating in our working life. Experimentation marks the point where we start to look at the change in a positive sense to try to work with it. New office layouts are fine-tuned, new systems are tried out to see what happens if I use it in certain ways, and new teams are engaged with.

As managers and promoters of change we need to remember to expect people's behavior to be affected as they come to terms with the change. We have to expect shock and denial, allowing people time, but giving them the information they need. We need to gradually introduce practical realities of the change—new office layouts, examples of the new computer transactions in order to help awareness of the reality that the change is going to happen. As soon as acceptance and experimentation start to occur, we must be ready to accelerate involvement in making the change. This could be meetings to plan details of the layouts, timings, or new definition of jobs—work that will help both the program and individuals in their understanding of the new way of working and their place in it.

We are likely to have known about the change for some time and have been through the curve ourselves. We need to allow others to do the same, at their own pace.

9.6 Willingness and Ability

Paul Hersey and Ken Blanchard originally developed the link between willingness and ability in relation to situational leadership. The approach shown in Figure 9.7 can be a useful way to move toward trying to gain consensus and approval for a proposal from a disparate group of people.

Within a population, we are likely to have people in all four of the boxes in Figure 9.7. As the ambassadors of change, the important thing for us is to recognize which box people are occupying and treat them accordingly.

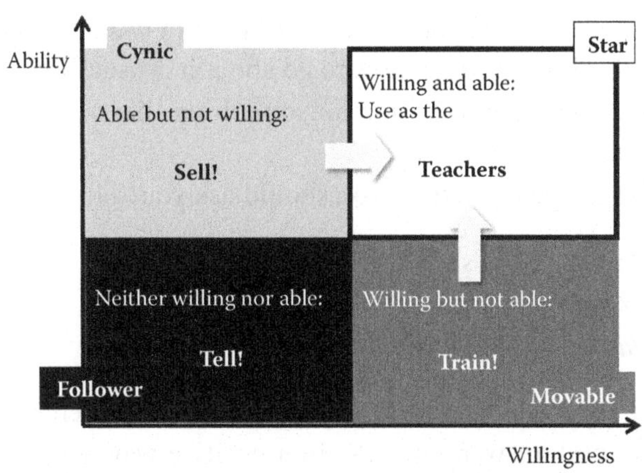

Figure 9.7 Willingness and ability.

Willing and able people understand our thinking and the need for change. They can be easily motivated to support change in their area. They are also able to assist, either because they have good skills, are in a senior position, or are just opinion leaders. They are the most useful of people, but sadly are rare. There may be only one (or even none) in a typical small meeting. When we have some willing and able people, we should build on their capability by asking them to undertake some part of the change—perhaps even lead a team including some willing but unable people.

Willing but not able people will be prepared to help, but unable to work out exactly how to proceed. They are usually good team members to assist the willing and able to undertake work or projects that are part of the change.

Unwilling and unable people are unlikely to be senior. They may cause difficulties, though. They sometimes seem like a heavy weight to shift. It is better to get these people occupied with something as useful as possible by giving clear instructions about what to do. This usually works by keeping them occupied and doing something useful to help the change.

Those who are *able but unwilling* are the most dangerous and more likely to derail the change. You need to get these people to change their view. Listening carefully to this group is important. First, they may have good points that highlight problems with our change and we may need to address these. Second, they may have a vested interest in keeping things as they are and we may be able to figure out what that is. It could even be that they put in place the current method we are trying to change. It could be that they will lose status or control because of the change. We need to quell their fears of the change and give them some benefits to look for. We need to sell the change to them. Do not try to use the willing and able people to persuade the unwilling and able. You know what may happen. Everyone could slip back into the unwilling side!

One of the best ways to sell is to use the structure described in the next section.

9.7 Presenting and Selling Your Case for Change

Whether you are going to put a case to a board or convince a workgroup or an individual, time spent thinking through how to go about it is usually time very well-spent. Not doing so risks the shock of rejection for something that may seem an obvious and sensible course of action to you.

Here are three suggested questions you should ask yourself:

1. *What is the problem*: Why should they do this?
2. *What is the proposal*: How might it impact them?
3. *What is the plan*: How will we do it?

These are three questions we try to answer when we present a consultancy case. If you can present good answers to these in a positive way, you should have a good chance of success:

9.7.1 What Is the Problem: Why Should They Do This?

Some of the people you need to approve your plans or take a lead role during implementation will perceive no benefits. Many people will be used to the status quo. They may like the way things are now. They may also see risks in your changes—perhaps to their own jobs. In any case, most people generally do not like change. The wastes in Lean recognize this and are used to help people to see the problems with current ways of working. To get people to accept change, you are going to have to convince them why they should.

Identifying and communicating benefits is the strongest approach to overcoming resistance. They may be about the good things the changes will bring, but they can also be about highlighting the negatives in the current situation. There has long been a saying in senior management: If there is no crisis—create one. The current situation may be too costly, causing us to lose business, or even physically unsafe, and so on. People need to be convinced that the current situation is poor and unacceptable. When presented with supportable facts along these lines, it is difficult to argue against the need for change.

Equally, your proposals could be used to gain advantage from a new situation: new technology that is applicable to your product, a new IT development, a new supplier, and so on. However, the attraction of new IT is not enough. Your benefit statement needs to be tangible, appeal to the whole audience, and be irresistibly attractive. Benefits need to be attractive to different audiences. Senior people may find increasing profits, growth, and shareholder value good reasons for change. These may also be attractive to teams of people working in the business. However, job security or better working conditions might be much more appealing benefits.

Data is my favorite way of convincing people of the need to change. Section 9.4 discusses how to use data to make a case that is difficult to dispute. Views and opinions about a problem will not do.

9.7.2 What Is the Proposal: How Might It Impact Them?

So many proposals are for a new piece of software which, apart from costing a lot, often seems to rely on the assumption that a miracle then occurs and all is well. Proposals often fail to explain how the proposal addresses the problem we have just defined. Many new software proposals seem to get away with this. I think this happens because someone else has implemented it, and since they are in a well-respected and successful business, it must be right.

A good proposal should address the root of the problem. A new way to do a particular job should show why and by how much it is better. From this it should be able to make a plausible estimate of how much better it will be. How much time is saved and by how much are the failures reduced or the sales increased? Plausibility is key to the task of convincing people that you really understand how to do something in a different, practical, and advantageous way. In the past, I have used all sorts of simulations to convince people that a new approach (often Kanban, VMI, P-cards, or similar) will work for them.

Some good guidelines for the proposal are

- It needs to be clear and plausible that the new way of working will solve all or most of the problems.
- Benefits need to be tangible, usually in terms of money saved or additional money made.
- People being introduced to the proposal should be able to see their place in it.
- If possible, the proposal needs to capture people's imagination as being exciting; this could be a new technology (but do not frighten people), or fashionable, or show success for the business and individuals.

9.7.3 What Is the Plan: How Will We Do It?

If the proposed solution is good, but unattainable because of cost, time-scale, or impracticality, we will create frustration. People might accept where we would like to be, but think there is no way of getting there. We need a step-by-step plan for implementation that can be seen to be achievable.

This is an implementation plan in its simplest form, but also has to address specific issues that people will see as stumbling blocks. Examples of this are, How will there be enough people to make the change and continue day-to-day business? Where will the investment money come from? How will we transfer all the data to the new system? How will we roll-out the change to such a large population; we can not do it in one hit, can we? My colleague Crispin Brown will not allow steps in a plan to be called "trials" because some people will see this as an invitation to prove they are a failure (pilots are OK). It is true, psychology can play a big part here. The plan has to address all of the headline issues which may otherwise cause people to conclude that it will not work. Just like the proposal, people need to be able to see their place in it (Figure 9.8).

Making change happen tools are probably the most valuable ones we have.

* * *

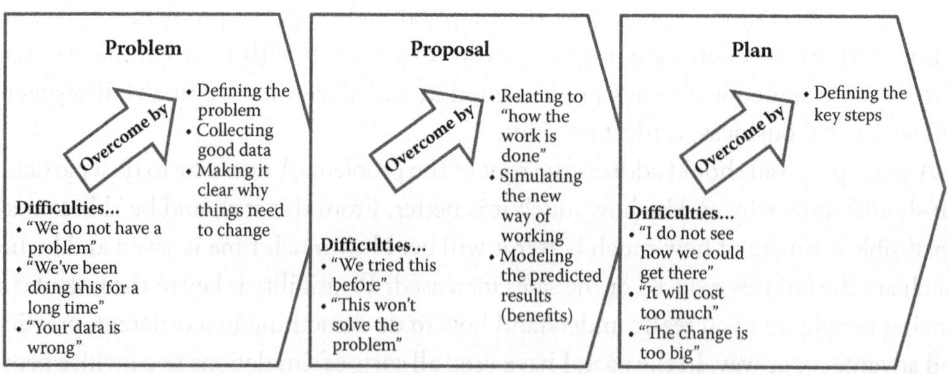

Figure 9.8 The case for change.

10

Product Design and Development and the Supply Chain

Few businesses make the entire product they sell. Most are reliant to a great extent on their supplier's manufacturing processes. The sourcing of the raw materials, components, assemblies, and services that are so important to the business is a great responsibility. Engineering and design are responsible for the product design, including that provided by suppliers. This is likely to include some new technology and some overseas supply as well as existing suppliers. Engineering and design therefore have a large part to play in sourcing.

A range of well-known research indicates that a large part of the supply base and how the supply chain will work is determined (and difficult to change) at the design stage. The supply base selected largely during the design and development therefore needs to be right for the operation of the supply chain and manufacturing, and so they also have a part to play in sourcing. The suppliers may also need to support product spares and maintenance long after product manufacture has ceased. Marketing and selling also have a legitimate input, both in terms of their input on salability of features and styling and the image that certain sourcing could have on product marketability. All of these areas cannot be in charge of sourcing. Mismanaged, this could be a recipe for disaster.

There are different ways to manage sourcing for new product designs. One business we know set up a complete team for each new product introduction in a separate building, including technical, marketing, manufacturing, and supply chain people. The program manager was responsible for all aspects of marketing, design, development, manufacture, and sourcing. The supply chain people on the team report to the program manager but have a dotted line responsibility to the central supply chain. In terms of responsibility, the function in overall charge of sourcing is less important. The important thing is strong management to ensure that proper input is taken into account.

New product design is the best and highly critical opportunity to introduce technical innovation into the product. Engineering should be at the forefront of the search for relevant new technology, but the supply chain also has a role to play. Existing suppliers are eager to present their innovations and supply chain people should actively engage engineering in reviewing them. Engineers will visit technology shows, and they should take supply chain people along when they go.

The current advances in 3-D metal printing are a good example of the need for integration between the engineer and the supplier. This new technology is so different that it requires a completely new approach to product design. The technology creates very thin layers of the component; one on top of another until the component is complete. Internal webs with voids can be created that would be very difficult with any other manufacturing technique. The alloy content can be varied in a specific part of the component, as can localized heat treatment. The design process, including how to take advantage of the strength, weight, and shapes possible, along with the selection of the right raw material powder, requires detailed understanding of the process. Work between the designer and supplier is therefore critical.

New areas of technology—3-D metal printing is a good current example—needs a dedicated team of engineers and supply chain people to investigate. It may be vital to the business to find and exploit new technology in terms of product features and producibility.

In a supply chain role I had in a company designing, manufacturing, and marketing electronic products, the supply chain was responsible for supplier selection. Several new product development teams ran concurrently and each had a supply chain representative. This representative coordinated the supply chain work on the project. New products always contained technical innovation and most required some new suppliers. Engineering took the lead in new technology, but supply chain evaluated the suppliers, researched alternatives, and managed the selection process. Areas of the product not requiring much technical innovation largely used existing suppliers. This cross-functional team was a good example of how all the different aspects of supplier selection can be managed.

Teamwork across supply chain, engineering, manufacturing, marketing, and service functions is essential to exploit supplier innovation.

* * *

Design for supply chain (DFSC) and design for manufacture (DFM) certainly highlight what we should try to do. We regularly encounter situations where huge cost savings are made when new designs are discussed between supplier and designer. These come from simple things like the repositioning of holes, changes to tolerancing, material specification, and thicknesses. However, it becomes more challenging when communication between supplier and designers is across different continents and different languages. In order to avoid the completion of a design that the supplier considers suboptimal, we need to set up a better communication routine than ever before. Teleconference facilities can be very helpful for regular meetings to share the objectives, review progress, and discuss the results of design reviews. Common CAE platforms become important, as well as common understanding of drawing standards, and even agreed-on e-mail protocols.

Design rules have been a standard way of working for electronic board manufacturing for as long as even I can remember. Some are better than others, but at least they give rules for dimensions and components that suit the manufacturer's standards.

Where application specific integrated circuits (ASICs) are designed by their customers, chip producers provide their own design tools for the customer to use. The criteria for supplier selection then needs to include the effectiveness of the design tools as well as the ability to provide the chips themselves.

Plastic moldings are usually designed by people who are well versed in issues such as mold flow and part extraction. Styling drawings may be done by the customer, or more likely an agency, but designs are usually done in conjunction with the final producer and the mold maker.

Machining is not quite the same. Equipment lists and general guidance on size capabilities are given, but how producible the design is depends on interaction between the machinists and the designer. There are great benefits from involving a machinist as the design develops—holes are positioned, tolerances are specified, and shapes are generally designed. Small differences in the design can make the component either easy (and cheaper) to make or nearly impossible. When machined parts are designed in Europe or the United States, for example, but will be manufactured in Asia, a raw material specification that can be met in Asian-produced stock is also important. Some suppliers have machine shops in low-labor-cost countries (LLCC) but prototype shops in Europe and the United States. The prototype shops have machines that are similar to those in the LLCCs and can give advice on DFM as well as producing small-volume parts.

A few design organizations, thankfully not too many, neglect the basics of standard sizes and components. When I first started working, we had no computer-aided design (CAD) or engineering (CAE), there were no personal computers either! Our standards were written on paper, and we had, for instance, a table of cap screw sizes. The manufacturer's standard diameter threads and lengths were marked. Also marked were those we already used in an existing product. If we wanted to use one in a new design, we used a standard size and preferably one we already used. The same applied to many other standard parts such as raw material, switches, and connectors. If your CAE system does not support this, you could revert to this old way of doing things—but using a spreadsheet approach on a shared drive, of course.

Design for supply chain can deliver great advantages but requires professionally organized communication between designers and suppliers.

* * *

Design teams should be supported by a member of the supply chain. I see them being responsible for the following:

1. Linking design teams with relevant activities underway, such as new technology initiatives with suppliers
2. Linking supply chain people to requirements such as the need for a new supplier or technology
3. Working with designers to understand and facilitate the use of new technology

4. Guiding designers when they need the input of suppliers (making sure they do not compromise the supply chain)
5. Facilitating DFM/DFSC by introducing supplier expertise to the design team
6. Representing the supply chain and bringing in other members of the team when specialist supply chain issues arise
7. Promoting the use of common and standard parts and materials
8. Getting prototype items quickly

Supply chain representatives on design programs may not be full-time, but they should have a desk colocated with the team. They will be far more aware of the project aims, difficulties, and time scales if they are colocated for at least some of their time. There is often a cultural divide between design teams and the supply chain if they only meet in scheduled meetings. Any proactive discussions between the design and supply chain teams will always bear fruit. If suppliers are involved there are likely to be even better results.

Supply chain representatives should support new product design proactively, supporting DFSC and supplier innovation.

* * *

Looking for new suppliers and technology is something that the supply chain should be proactively involved in doing. Most businesses are at least to some extent reliant on supplier development of technology—some do styling, sales, and marketing and rely on their suppliers to develop new technology, detailed designs, and production. Although engineers will be involved at the forefront of spotting new technologies and suppliers, the supply chain equally should be. Commodity groups (e.g., for electronics or display screens) should have new technology road maps as an area of focus. The supply chain should also have a close relationship with engineers in general in order to have new technology discussions with current and potential suppliers, visit technology shows with the engineers, read the right technology magazines, and so forth. The supply chain should also enable potential new technology suppliers to approach them through channels such as their website. Potential new suppliers are asked to sign an on-line NDA. Following this, the supplier is provided with an on-line facility to enter a set of standard information, followed by a free-format area for him to propose what he can supply. The supply chain, together with engineering can then decide if the supplier has potential and whether to take the discussions further.

Suppliers contribute increasingly to design and new technology—and the supply chain needs to be proactively involved in finding it.

* * *

10.1 Supplier Involvement in Product Design Teams

Suppliers are often the best equipped to do the detailed design work on the product they will produce for us. For example:

- On plastic moldings, styling drawings can be developed into manufacturing drawings of parts that will enable good flow, appropriate draft angles, and can be ejected well. Plastic molders offer this service or do so through specialist design partners.
- In-house electronics design teams frequently use customized chips that will be manufactured to their own design. However, they will use design tools that are supplied by the chip manufacturer. To a large extent, the supplier of the chips cannot be changed once the chip design is done using his or her design tools. Supplier selection will in this case need to consider the quality of the design tools as well as chip supply and future pricing and price-reduction curves.
- On a very large scale, automotive manufacturers now use multiproduct design platforms that include major supplied parts. Engines and bodies are likely to be made in-house. However, products such as electronic control systems and lighting systems will usually come from a partner supplier. The partners will certainly be involved in the design work for the model, but increasingly they will be the designer of their equipment for a range of models or platforms.
- There are also a number of very capable design engineering companies who will carry out the final design detailing. This is often cheaper because it can be done in an LLCC. These organizations are generally very good at the detailing work, adherence to standards and DFM/DFSC.

Suppliers are producing an increasing proportion of our product and are therefore naturally part of the design effort. The selection of suppliers has always needed to include a range of business functions. Now, suppliers often need to be selected for more than their product supply capabilities. Selection also needs to take place far earlier in the product creating process.

Appendix: Example Supply Chain Role Profiles

- Vice president, supply chain
- Commodity group leader, electronic components
- Senior buyer, raw materials
- Buyer, indirect and consumables
- Vendor scheduler, packaging
- Sales and operations planning manager

EXAMPLE ROLE PROFILE

TITLE: Vice President, Supply Chain

REPORTS TO: Chief Executive Officer

PURPOSE: To develop and optimize the contribution made by supply chain to the business in terms of flexibility, service levels, technical contribution, and total costs

KEY RESULT AREAS:

- To develop the ability of the supply chain organization to interface with technical and customer facing teams in order to deliver improved technical contribution and service levels from the supply base
- To develop the supply base in order to deliver optimum technical input and service levels at optimum cost on a global basis
- To manage the assessment and mitigation of major supply chain risks to the business
- To improve supply chain and associated processes to reduce administrative cost and improve response
- To agree on the direction, investment, and development of supply base capabilities with the board

KEY MEASURES:

- Cost reductions in direct and in-direct purchases
- Project stage delays due to technical developments from supply base
- Project stage delays due to supply chain/supply base lead-times
- No major supply chain risks unmitigated on major projects
- Reduction in lead time/project stages for supply chain critical path activities
- Creation, agreement with board, and delivery against supply chain development plan

PERSON PROFILE

KNOWLEDGE AND EXPERIENCE

- Senior-level track record and exemplary supply chain contributions to large international engineering programs
- Appreciation of technical issues and requirements in aerospace
- Knowledge of and experience in global supply chain best practice

SKILLS	BEHAVIORS
Communication and interpersonal skills at senior level	High drive
Commercially astute	Change-orientated
Leadership and motivational skills	Tenacious
Emotional intelligence	Goal-orientated
	Self-confidence
	Extrovert

EXAMPLE ROLE PROFILE

TITLE: Commodity Group Leader—Electronic Components

REPORTS TO: Vice President, Supply Chain

PURPOSE: To manage and develop the use of global suppliers for electronic components to optimize product development timescales and costs, supply lead times, and costs at lowest risk for the business

KEY RESULT AREAS:

- To manage the global sourcing program, working in conjunction with technical teams ensuring the best support for technical development and optimum service levels and costs for manufacture
- To work with technical staff to optimize exploitation of technical and industry developments to enhance product performance and project development time scales and deliver world-class service levels to manufacturing
- To develop and maintain strong and appropriate relationships with key electronic suppliers
- To agree on and manage appropriate contractual terms and conditions with key electronic suppliers in order to maximize business potential and minimize risk
- To set up and monitor arrangements for delivery schedules that achieve high service levels to manufacturing and low inventory risk
- To manage the electronic component commodity group team to optimize their effectiveness and personal development potential
- To measure and report the performance of the electronics commodity group against agreed-on targets

KEY MEASURES:

- Reduction in supplier lead time for technical development, tooling, and product introduction for key ASICs
- Reduction in supplier lead time and increased schedule flexibility for component supply
- Level of total cost savings achieved across electronic components
- No level 1 risks unmitigated in electronic component supply
- Quality failures for electronic parts from in-house test and during product life

PERSON PROFILE

KNOWLEDGE AND EXPERIENCE

- Knowledge of and experience in global procurement best practice
- Experience in the computer/electronics industry
- Appreciation of electronics technical issues and current developments
- Knowledge of and experience in commercial contracting on an international basis

- Knowledge of and experience in team leadership and people management
- Knowledge of and experience in trade compliance and control issues
- Degree-level education

SKILLS	BEHAVIORS
Commercially astute	High drive
Communication skills	Tenacious
Negotiation	Team leader
Commercial contracts	Goal-orientated
Analytical	Self-confidence
Interpersonal	
Emotional intelligence	

EXAMPLE ROLE PROFILE

Note: The old view of the buyer who sourced, negotiated, and placed orders has been replaced by one where a broad understanding of the business needs and processes, together with good communication skills, is needed.

TITLE: Senior Buyer, Raw Materials
REPORTS TO: Metal manufacture commodity group manager
PURPOSE: To source and manage the procurement of raw materials for internal and subcontract manufacture for all global sites
KEY RESULT AREAS:

- To source and develop suppliers for a range of raw materials, but mainly steel and copper in medium volumes, optimizing quality, cost, and delivery
- To monitor movement and the drivers of key raw material prices and agree on and manage hedging, forward buying, and other arrangements to reduce the risk of price instability on product costs
- To negotiate contractual terms and conditions for product supply including hedge and buy-forward options
- To agree on and manage the raw material procurement processes for buyers procuring raw materials in all global sites
- To measure and report on the performance of raw material supply in terms of price, delivery lead time, and quality

KEY MEASURES:

- Cost of raw materials in comparison with global indices
- Raw material lead time and delivery for all global sites
- Raw material quality for all global sites

PERSON PROFILE

KNOWLEDGE AND EXPERIENCE

- Knowledge and experience of global raw material procurement, particularly steel and copper
- Experience of manufacturing for large machines and equipment ideally related to mining industry equipment
- Appreciation of technical issues related to raw material alloy content and machining processes

- Knowledge and experience of commercial contracting on international basis
- Knowledge and experience of team leadership and people management
- Degree-level education

SKILLS	BEHAVIORS
Commercially astute	High drive
Communication skills	Communicating
Negotiation	Goal-orientated
International commercial contracts	Team leadership
Analytical	Self-confidence
Interpersonal	

EXAMPLE ROLE PROFILE

TITLE: Buyer, Indirect and Consumables

REPORTS TO: General commodity group manager

PURPOSE: To source and manage the procurement and supply processes and suppliers for indirect and consumable products and materials for the U.K. sites

KEY RESULT AREAS:

- To source and develop suppliers for a wide range of regular and irregular indirect and consumable products, optimizing quality, cost, and delivery
- To set up administratively economical yet low-risk and controlled processes for procurement and supply of a wide range of indirect products and materials for production and office use
- To agree on contractual terms and conditions for product supply
- To measure and report on the performance of indirect and consumable products procurement and drive improvements against agreed-on targets

KEY MEASURES:

- Reduction in total cost of supply for indirect and consumable products
- Reduction in administrative lead time and costs for procurement and supply of indirect and consumable products
- Surveyed internal customer satisfaction

PERSON PROFILE

KNOWLEDGE AND EXPERIENCE

- Knowledge of and experience in global procurement best practice
- Knowledge of and experience in commercial contracting on an international basis
- Knowledge of and experience in team leadership and people management
- Degree-level education

SKILLS	BEHAVIORS
Commercially astute	High drive
Communication skills	Goal-orientated
Negotiation	Self-confidence
Commercial contracts	
Analytical	
Interpersonal	
Emotional intelligence	

EXAMPLE ROLE PROFILE

Note: The person in this type of role manages the scheduling of deliveries, usually for one site and against purchase orders that have been put in place by some kind of buyer. Reporting can be into a commodity group or into a planning group—often connected to the requirements of one site, and sometimes reporting directly into the manufacturing organization.

TITLE: Vendor Scheduler–Packaging
REPORTS TO: Master scheduler
PURPOSE: To manage the scheduling and delivery of packaging materials to line-side in the plant
KEY RESULT AREAS:

- To source and develop suppliers for a wide range of regular and irregular indirect and consumable products, optimizing quality, cost, and delivery
- To set up administratively economical yet low-risk and controlled processes for procurement and a supply of a wide range of indirect products and materials for production and office use
- To agree on the contractual terms and conditions for product supply
- To measure and report the performance of indirect and consumable products' procurement and drive improvements against agreed targets

KEY MEASURES:

- No schedule delays due to nonavailability of packaging
- Inventory level of packaging
- Obsolescence costs of packaging
- Emergency shipment costs

PERSON PROFILE
KNOWLEDGE AND EXPERIENCE

- Appreciation of the manufacturing processes for cartons, cans, boxes, and printing
- Experience of food packaging
- Knowledge of modern supply processes such as vendor-managed inventory

SKILLS	BEHAVIORS
Communication skills	High drive
Numeracy	Goal oriented
Analytical	Team member
Interpersonal	Self-confidence

EXAMPLE ROLE PROFILE

Note: This role varies in detail and seniority and is sometimes called the master scheduler or the chief planner. The role and the process (see section on sales and operations planning [S&OP]) are critical in most businesses because they help senior people manage the risks and imbalances between *demand* and *supply*.

TITLE: Sales and Operations Planning Manager
REPORTS TO: Chief operating officer
PURPOSE: To manage the S&OP process to optimize the business response to changing customer demand
KEY RESULT AREAS:

- To work with sales and marketing to manage the process and assess the nature and implications of changing forecast and actual customer demand
- To work with the supply chain and manufacturing teams to assess the capability to adjust supply volumes and timescales
- To manage the S&OP process to propose updated master schedule plans to the S&OP approval team
- To propose master schedule updates that optimize the business capability to respond to customer demand
- To propose appropriate risk management strategies and investments to the S&OP team
- To present plan revisions and recommendations at the S&OP meeting
- To record plan amendments and other actions that are agreed at the S&OP meeting and publish them
- To work with the supply chain and manufacturing teams to implement the S&OP meeting agreed output plans and actions
- To measure and report S&OP performance in terms of adherence to plan

KEY MEASURES:

- Adherence to S&OP plan
- Inventory level of finished goods and contingency items
- Short lead-time expediting costs
- S&OP process timetable adherence

PERSON PROFILE
KNOWLEDGE AND EXPERIENCE

- Appreciation of drivers of volatility in demand and finite ability to adjust supply
- Knowledge and experience of capital goods manufacture
- Knowledge and experience of enterprise resource planning (ERP) systems, especially material requirements planning (MRP)
- Knowledge of the S&OP process

SKILLS	BEHAVIORS
Communication skills	High drive
Numeracy	Attention to detail
Analytical	Patience
Systems understanding	Self-confidence

Suggested Further Reading

Gavin Kennedy, *Everything Is Negotiable*, Random House, 2008, ISBN: 9781847940018.

Gerard I. Nierenberg, *The Art of Negotiating*, Barnes Noble, 1995, ISBN: 978-1566198165.

James P. Womack, Daniel T. Jones, and Daniel Roos, *The Machine That Changed the World*, ISBN-13: 978-0-7432-9979-4.

Kenneth H. Blanchard and Spencer Johnson, *The One Minute Manager*, ISBN: 978-0-00-636753-6.

Michael Hammer and James Champy, *Reengineering the Corporation*, Harper Business, 1993, ISBN13: 9780060559533.

Oliver W. Wight, *Manufacturing Resource Planning: MRP II: Unlocking America's Productivity Challenge*, 1984, ISBN: 13 9780939246038.

Purchasing Must Become Supply Management, *Harvard Business Review*, September 1983.

Roger Fisher and William Ury, *Getting to Yes*, Houghton Mifflin, 1981, ISBN: 01401.57352.

Index

D

I

J

K

L

Q

R

S

T

U

www.ingramcontent.com/pod-product-compliance
Lightning Source LLC
LaVergne TN
LVHW082025200926
848481LV00010B/2925
* 9 7 8 1 7 6 4 6 4 6 2 5 3 *